7th International Conference on Global Optimization and Its Application (ICoGOIA 2018)

Get Fresh Insights into the World of Global Optimization and Green Science & Technology for Collaboration Sustainability

IOP Conference Series: Materials Science and Engineering
Volume 621

Bali, Indonesia
30 - 31 August 2018

Editors:

Eneng Tita Tosida
Fajar Delli Wihartiko

ISBN: 978-1-7138-0901-2
ISSN: 1757-8981

Printed from e-media with permission by:

Curran Associates, Inc.
57 Morehouse Lane
Red Hook, NY 12571

Some format issues inherent in the e-media version may also appear in this print version.

Institute of Physics
Dirac House, Temple Back
Bristol BS1 6BE UK

Phone: 44 1 17 929 7481
Fax: 44 1 17 920 0979

techtracking@iop.org

Additional copies of this publication are available from:

Curran Associates, Inc.
57 Morehouse Lane
Red Hook, NY 12571 USA
Phone: 845-758-0400
Fax: 845-758-2633
Email: curran@proceedings.com
Web: www.proceedings.com

TABLE OF CONTENTS

IOP Conf. Series: Materials Science and Engineering **621** (2019) 011001 doi:10.1088/1757-899X/621/1/011001

IOP Conf. Series: Materials Science and Engineering **621** (2019) 011001 doi:10.1088/1757-899X/621/1/011001

ICoGOIA

The 7th International Conference
on Global Optimization and Its Application 2018

The Conference

Optimization is a process to achieve the ideal or optimal value of an objective function. In the discipline of mathematics, optimization also refers to the study of problems trying to find the minimum and maximum values of a real function. To achieve these minimum and maximum values, some processes were systematically carried out by selecting integers or real numbers that will provide optimal solutions.

The true discriminant between "easy" and "hard" optimization problems in mathematical programming is the convex/non-convex issue. A problem is convex if it is a minimization of a convex function (or a maximization of a concave function) where the admissible points are in a convex set. The fundamental result in convex analysis says that a locally optimal solution of a convex problem is also globally optimal. This is practically very important since algorithmic termination tests occur at each iteration in the algorithm, they must be computationally very efficient; thus, virtually all termination conditions of optimization algorithms are local conditions, i.e. they test whether the current solution is locally optimal with respect to a pre-defined neighborhood. If the problem is convex this is enough to guarantee that the solution is globally optimal. Non-convex problems, on the other hand, may have many different local optimal, and choosing the best one is an extremely hard task.

The field of applied mathematics that studies extremal locations of non-convex functions subject to (possibly) non-convex constraints is called Global Optimization. Research on global optimization has been continuing rapidly and many of the findings are of new techniques that are more sophisticated and efficient.

The 7th International Conference on Global Optimization and Its Application (ICoGOIA) 2018 is hosted and organized by Faculty of Mathematics and Natural Sciences Universitas Pakuan, Indonesia.

Conference Publication

All papers received are reviewed by two peer reviewers and selected papers will be published in IOP Conference Series: Materials Science and Engineering, indexed by SCOPUS.

IOP Conf. Series: Materials Science and Engineering **621** (2019) 011001 doi:10.1088/1757-899X/621/1/011001

Conference Theme

The theme of this year conference is:

Get Fresh Insights into the World of Global Optimization and Green Science & Technology for Collaboration Sustainability.

Organization

The conference is organized by:

Faculty of Mathematics and Natural Sciences, Universitas Pakuan, Jl. Pakuan PO Box 452, Bogor 16143, http://ilkom.unpak.ac.id/

Supported by:
1. Indonesian Operations Research Association (IORA)
2. Universitas Padjadjaran, Indonesia
3. Osaka Prefecture University, Japan
4. Universiti Sultan Zainal Abidin, Malaysia
5. Universiti Tun Hussein Onn, Malaysia
6. STIKOM Bali, Indonesia
7. Universitas Sumatra Utara, Indonesia
8. Universitas Sam Ratulangi, Indonesia
9. Institut Pertanian Bogor, Indonesia
10. Vietnam National University, Vietnam
11. Universitas Terbuka, Indonesia
12. Universiti Malaysia Pahang, Malaysia

Participants

International Conference on Global Optimization and its Applications 2018 (ICoGOIA 2018) was attended by about 113 participants from 10 countries and 20 institutions/universities, including but not limited to academicians (professors, lecturers, undergraduate and graduate students, high school teachers) and researchers in mathematics, statistics, economics, finance, management, operations research, applied science and related fields.

Committee

Conference Chairs

Prasetyorini (Universitas Pakuan, Indonesia)
Sri Setyaningsih (Universitas Pakuan, Indonesia)

Program Chairs

Eneng Tita Tosida (Universitas Pakuan, Indonesia)
Prihastuti Harsani (Universitas Pakuan, Indonesia)
Diana Widiastuti (Universitas Pakuan, Indonesia)

IOP Conf. Series: Materials Science and Engineering **621** (2019) 011001 doi:10.1088/1757-899X/621/1/011001

Diana Rachmawati (Universitas Padjadjaran, Indonesia)
Rohaizan (Universiti Tun Hussein Onn, Malaysia)
Subiyanto (Universitas Padjadjaran, Indonesia)
Endang Suryana (Universitas Padjadjaran, Indonesia)
Pramono Sidi (Universitas Terbuka, Indonesia)
Moerfiah (Universitas Pakuan, Indonesia)
Arie Qurania (Universitas Pakuan, Indonesia)
Fajar Delli Wihartiko (Universitas Pakuan, Indonesia)
Agung Prajuhana (Universitas Pakuan, Indonesia)
M. Saad Nurul Ishlah (Universitas Pakuan, Indonesia)
Dedy Panji Agustino (Universitas Pakuan, Indonesia)

Advisory Committee

Sudradjat Supian (Universitas Padjadjaran, Indonesia)
Preda Vasile (University of Bucharest, Romania)
Mustafa Bin Mamat (Universiti Sultan Zainal Abidin, Malaysia)
Abdul Talib Bon (Universiti Tun Hussein Onn, Malaysia)
Abby Tan Cee Hong, PhD. (Universiti Brunei Darussalam, Brunei)
Muhammad Zarlis (Universitas Sumatra Utara, Indonesia)
Benny Pinontoan (Universitas Sam Ratulangi, Indonesia)
Bermawi P. Iskandar (Institut Teknologi Bandung, Indonesia)
Abdul Hakim Halim (Institut Teknologi Bandung, Indonesia)
Soewarto Hardhienata (Universitas Pakuan, Indonesia)
Toni Bakhtiar (Institut Pertanian Bogor, Indonesia)
A. Beny Mutiara (Universitas Gunadharma, Indonesia)
Asep K. Supriatna (Universitas Padjadjaran, Indonesia)
Edy Soewono (Institut Teknologi Bandung, Indonesia)
Hadi Sutanto (Universitas Katolik Indonesia Atma Jaya, Indonesia)
Herman Mawengkang (Universitas Sumatra Utara, Indonesia)
Widodo (Universitas Gadjah Mada, Indonesia)
Rina Indiastuti (Universitas Padjadjaran, Indonesia)
Bibin Rubini (Universitas Pakuan, Indonesia)
Admi Syarif (Universitas Lampung, Indonesia)
Amril Aman (Institut Pertanian Bogor, Indonesia)
Sukono (Universitas Padjadjaran, Indonesia)
Dadang Suherman (STIKOM Bali, Indonesia)
Sri Hariyati (Universitas Terbuka, Indonesia)
Syafrizal (Universitas Sultan Ageng Tirtayasa, Indonesia)
Fatimah binti Mahmud (Universiti Malaysia Pahang, Malaysia)

Reviewers

Sudradjat Supian, M.Sc (Universitas Padjadjaran, Indonesia)
Mustafa Bin Mamat (Universiti Sultan Zainal Abidin, Malaysia)
Abdul Talib Bon (Universiti Tun Hussein Onn, Malaysia)
Muhammad Zarlis, M.Sc (Universitas Sumatra Utara, Indonesia)
Benny Pinontoan (Universitas Sam Ratulangi, Indonesia)
Soewarto Hardhienata (Universitas Pakuan, Indonesia)
Toni Bakhtiar (Institut Pertanian Bogor, Indonesia)
A. Beny Mutiara (Universitas Gunadharma, Indonesia)
Sukono (Universitas Padjadjaran, Indonesia)
Hafizan Juhair (Universiti Sultan Zainal Abidin, Malaysia)
Fatma Susilawati Mohamad (Universiti Sultan Zainal Abidin, Malaysia)

IOP Conf. Series: Materials Science and Engineering **621** (2019) 011001 doi:10.1088/1757-899X/621/1/011001

Secretariat

Faculty of Mathematics and Natural Sciences, Universitas Pakuan,
Jl. Pakuan PO Box 452, Bogor 16143, Indonesia

Phone/Fax	:	+62 251 8375547
Email	:	icogoia@unpak.ac.id
Website	:	http://icogoia2018.fmipa.unpak.ac.id/
Contact Persons	:	Diana Widiastuti (+62 81383931071) Eneng Tita Tosida (+62 8129599049) Prihastuti Harsani (+62 81383639139) Rohaizan (+60 199399019)

IOP Conf. Series: Materials Science and Engineering **621** (2019) 011001 doi:10.1088/1757-899X/621/1/011001

The 7th International Conference
on Global Optimization and Its Application 2018

Welcoming Speech

Dr. Prasetyorini, *Conference Chair*

Prof. Sudradjat Supian, president of the Indonesian Operations Research Association and dean of FMIPA University of Padjadjaran, Prof. Hiroshi Tsuji, president of Osaka Prefecture University Japan, Dr. Bibin Rubini, Rector of Universitas Pakuan or his representative, Dr. Pham Van Hung, Researcher and Lecturer of National Vietnam University, Dr. Arief Liehman, Deputy Director Monash Energy Materials and System Institute, Monash University Melbourne Australia, distinguished guests, ladies and gentlemen.

On behalf of the Conference Committee, I would like to welcome all attendees and thank you for your presence in this conference. In particular we would like to thank the Rector of Universitas Pakuan for their willingness to give a speech and at the same time officially open this conference. Thank you also to the Keynote Speakers and the speakers and conference participants who have attended this conference.

The holding of this conference aims to gather researchers, lecturers, and observers of science and technology to gather together to share knowledge and experience in optimizing activities in various fields. Therefore, this conference took the theme "Optimization of Green Science and Technology for Collaboration Sustainability." This international conference is a series of conference activities that have been carried out previously and the conference that we held in 2018 is a conference for the seventh time. Therefore, we call it "The 7th ICoGOIA 2018".

Distinguished ladies and gentlemen.

We report that this conference will be attended by 113 participants from 10 countries and 20 institutions/universities. There are the same number of papers which will be presented and discussed in this conference. Some of the papers discussed in this conference will be published in Scopus indexed IOP Publications and parts of the others will be published in proceedings from this international conference.

In this good opportunity, once again we would like to thank the Rector of Universitas Pakuan for their support, the speakers for the contribution of the paper, and thank you also to the steering committee and the organizing committee so that the 7th Icogoia conference 2018 was held at Discovery Kartika Hotel Denpasar Bali Indonesia can be carried out properly.

IOP Conf. Series: Materials Science and Engineering **621** (2019) 011001 doi:10.1088/1757-899X/621/1/011001

Do not forget we remind you that in addition to following the ICoGOIA conference we also recommend you to explore the island of Bali to enjoy the foods and its natural beauty. Have a nice conference and pleasant stay at Bali.

Welcoming Speech

Dr. Bibin Rubini, *Rector of Universitas Pakuan*

Prof. Sudradjat Supian, president of the Indonesian Operations Research Association and dean of FMIPA University of Padjadjaran, Prof. Hiroshi Tsuji, president of Osaka Prefecture University Japan, Dr. Pham Van Hung, Researcher and Lecturer of National Vietnam University, Dr. Arief Liehman, Deputy Director Monash Energy Materials and System Institute, Monash University Melbourne Australia, distinguished guests, ladies and gentlemen.

It is a great honor for me to welcome all of you to the International Conference on Global Optimization and its Applications 2018 (ICoGOIA 2018), this time in Bali. On behalf of Universitas Pakuan, as the organizing committee supported by IORA and other university partners, I would like to express my heartfelt gratitude for your presence in this opening ceremony of the conference.

I am particularly thankful to Prof. Sudradjat from Indonesia, Prof. Hiroshi Tsuji from Japan, Dr. Pham Van Hung from Vietnam, and Dr. Ariel Liebman from Australia for taking time off from your busy schedule to deliver keynote speech in this important conference. Therefore, I am especially honored to welcome you again here, in this event.

I very much appreciate the initiative of Prof. Sudradjat Supian from University of Padjadjaran Indonesia, Prof. Muhamad Zarlis and Prof Herman Mawengkang from Sumatra Utara University, Prof. Benny Pinontoan from Universitas Sam Ratulangi Manado, Prof. Soewarto Hardhienata from Universitas Pakuan, Prof. Abdul Talib bin Bon from UTHM Malaysia in establishing and coordinating the conference.

We know that to organize an international conference is needed a hard works and many preparations. But I do believe that in the near future ICoGOIA will be able to promote the increase of the publication from many researchers and university lectures. This event can be considered as good platforms for researchers, academicians and professionals in Indonesia and also researchers, academicians and professionals from other countries to meet each other, exchange ideas and strengthen their collaboration in the field of science and technology.

I would also like to thank Dr. Prasetyorini the Dean of Mathematic and Natural Sciences (FMIPA) of Universitas Pakuan and all members of Organizing Committee who have worked very hard so that this conference can be held properly.

Finally, I wish all participants enjoy a most fruitful day with interesting and stimulating discussions and exchange of knowledge in this conference.

Thank you very much.

IOP Conf. Series: Materials Science and Engineering **621** (2019) 011001 doi:10.1088/1757-899X/621/1/011001

ICoGOIA

The 7th International Conference
on Global Optimization and Its Application 2018

Table of Contents

No	Authors	Title
13	C H Lien, S Vaidyanathan, A Sambas, S Sampath, Sukono, M Mamat	*A new 3-D chaotic system with four quadratic nonlinear terms, its global chaos control via passive control method and circuit design*
14	L G Dolvis, S Vaidyanathan, K Jacques, A Sambas, Sukono, M Mamat	*A new 4-D hyperchaotic system with four-scroll hidden attractor, its properties and bifurcation analysis*
15	Sukono, B Subartini, P Thalia, S Supian, E Lesmana, R Budiono, H Juahir	*Modeling the impacts of constant price GDP and population on CO2 emissions using Cobb-Douglas model and ant colony optimization algorithm*
16	Y Hidayat, S Winarni, T Purwandari, A Sukmawati, Sukono	*Comparative analysis of inline quality control system process capability with offline lab QC against product specification by MSPC method*
17	I G N M Jaya, Y Andriyana, B Tantular, Zulhanif, BN Ruchjana	*Local spatiotemporal Moran's index for dengue disease clustering*
18	M D Johansyah, H Napitupulu, E Djauhari, Sukono, J Nahar	*Numerical comparison of some crossover in genetic algorithm for two dimensional global minimization problem using C++*
19	H Juahir, A Ghazali, A Ismail, M Mohamad, F M Hamzah, S Sudianto, M L Lasim, M A Shahriz	*Spatio-temporal water quality assessment of Danau Kota Lake using environmetric techniques*
20	R Budiono, H Juahir, M Mamat, Sukono, S Supian, M Nurzaman	*Modeling and analysis of CO2 emissions in million tons of sectoral greenhouse gases in Indonesia*
21	I M Sulaiman, Sukono, S Supian, M Mamat	*New class of hybrid conjugate gradient coefficients with guaranteed descent and efficient line search*

IOP Conf. Series: Materials Science and Engineering **621** (2019) 011002 doi:10.1088/1757-899X/621/1/011002

Peer review statement

All papers published in this volume of *IOP Conference Series: Materials Science and Engineering* have been peer reviewed through processes administered by the proceedings Editors. Reviews were conducted by expert referees to the professional and scientific standards expected of a proceedings journal published by IOP Publishing.

IOP Conf. Series: Materials Science and Engineering **621** (2019) 012001 doi:10.1088/1757-899X/621/1/012001

Wireless Sensor Network Server for Smart Agriculture Optimatization

[1]Deden Ardiansyah, [2]Akbar Sugih Miftahul Huda, [3]Darusman, Ramandha [4]Gilang Pratama, [5]Andika Pratama Putra
[1,2,3]Jl. Ciheuleut Pakuan PO Box 452 Bogor 16143 West Java, Indonesia

[1]ardiansyahzhigadeden@gmail.com, [2]akbar.sugih@gmail.com, [3]darusman989@gmail.com, [4]rama.gilangpratama@gmail.com, [5]pratamaputraandika93@gmail.com

Abstract: Server wireless sensor network (WSN) is a server that can retrieve data from wireless sensor network so that server can give service like normal server to be accessed by all client whenever and wherever. WSN Server to facilitate smart farming requires a server to serve all the needs of agricultural data so that the data can be processed and optimized for the needs of smart farming. wireless sensor network server serves as a server and can receive data from multiple sensors through WSN. Then in the save to the database in real time, then visualized into the form of the website and can be accessed via the Internet network.

Keyword: WSN server, IoT, Agriculture, Smart farming, Moisture sensor

1. Introduction

In agriculture, soil moisture has a major role in the process of the plant growth. This is because the pant needed water as a natural source for its's growth. The type of the paddy that can be known by the soil moisture in the farming area [1]. The moisture of the farm field can be known by checking it directly. The process to measure the moisture and the temperature of the farm field can be conducted by using thermometer and hygrometer. This process can consume a lot of time and effort to know the exact moisture of the farm field. The farmer need an easier way to know the condition of the farm, since the farmer usually has a huge area of farm field. One of the method to know the condition of the farm field is by putting a sensor in the farm field to monitor the condition from distance. One of the network that can receive data from the sensor on the farm field and transmit it wirelessly is wireless sensor network.

Wireless sensor network server (WSN) is a server that can retrieve data from wireless sensor network so that server can give service like normal server to be accessed by all client whenever and wherever. WSN Server to facilitate smart farming requires a server to serve all the needs of agricultural data so that the data can be processed and optimized for the needs of smart farming. Wireless sensor network server serves as a server and can receive data from multiple sensors through WSN. Then in the save to the database in real time, then visualized into the form of the website and can be accessed via the Internet network.

Wireless sensor network (WSN) is a network system that connect sensors and equipped with communication devices to receive information or data from devices to WSN server. WSN usually used to monitored agricultural or plantation environment. WSN can even use to monitor disaster event. Wireless sensor network is widely used in agricultural domain for efficient farming management. The application of the network system for agricultural environment monitoring was successfully developed in this research[2].

Optimum soil moisture for rice cultivation with System of Rice Intensification (SRI) is very important with the aim to increase rice production and water productivity. Soil moisture is classified into three levels based on the water retention curve, wet, slightly wet (medium) and dry. Soil moisture at the wet level for the initial and vegetative phases is very important for plants, especially the rooting areas in providing sufficient water for root, stem and leaf growth[3].

IOP Conf. Series: Materials Science and Engineering **621** (2019) 012001 doi:10.1088/1757-899X/621/1/012001

One of the monitoring system in agricultural is to used WSN to monitored the environment from distance. This system can be manifested by designing a WSN server. This WSN server used to collect data from sensors trough WSN and stored in the database server. This data can be seen remotely by web.

2. Methodology

The methodology that used in this research is network development life cycle can be seen in Figure 1. Figure 1 shows the phase of the methodology used in this research. The methodology consists of analysis, design, simulation prototyping, implementation, monitoring and management. The description of each phase are explain as follow[4]

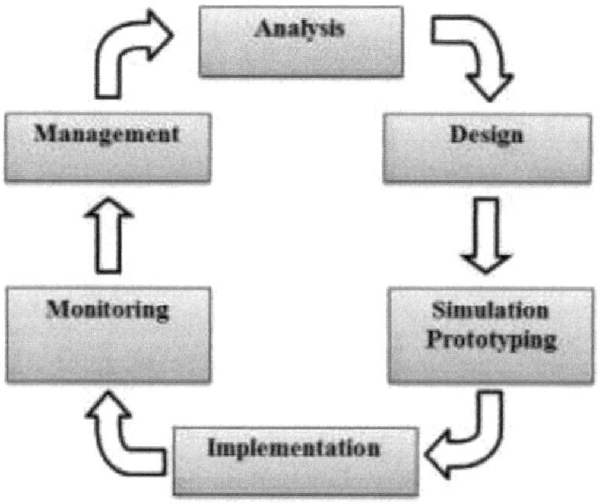

Figure 1. Network development life cycle

2.1. Analysis

The problem in this research were analyzed and to measure the requirement needed. The problem was the connectivity between the database and the microcontroller were unavailable. The WSN for smart farming optimization can be used to monitor the farm from the distance and able to store the data into the database.

2.2. Design

The network topology design was created. By creating the design of the network topology and the interconnection between the database and the sensor the system can provide reliable data from the farm. The design created were include the network topology, the data access design and cabling.

2.3. Simulation

After the analysis and the design phase, the next phase is to simulate using cisco packet tracer to test the readiness of the system for implementation

2.4. Implementation

In the implementation phase everything that has been done were implemented. The implementation phase will determine the result weather a failure or a success. The implementation phase was divide as follow:

1. The sensor process while activated produced the output as data to send to the database
2. Wireless router as communication media from sensors
3. The server computer will receive the data through WSN to store into the database and can be viewed via website

2.5. Monitoring

IOP Conf. Series: Materials Science and Engineering **621** (2019) 012001 doi:10.1088/1757-899X/621/1/012001

The monitoring phase is to see the reliability condition of the system, the flow of the transmitted packet in the network and the stability of the connection.

2.6. Management

The management phase is to manage the network so the WSN communication among all devices can be maintain.

3. Result

The condition of the farm field can be known by visiting the farm directly and measure the moisture using a measuring device such as hygrometer. This activity takes time and to know the moisture of the farm field. To ease the process to know the condition of the farm field, the WSN were needed so the condition of the farm field can be seen remotely. By implementing the WSN in the farm field the farmer can be able to measure the moisture of their farm field.

3.1. Analysis

The research uses sensors to measure the moisture of the farm field and an Internet of Things (IoT) module to transfer the data from the sensors to the WSN server.

3.1.1. Moisture sensor and Architecture of Sensor Node

The sensor used is to measure the soil moisture which is YL 69. The voltage that the sensor output changes accordingly to the water content in the soil. The voltage increase when the soil is wet and the voltage decrease when the soil is dry. The sensor works at 5v voltage with 10 bits of digital data which means can produce a digital value from 0 to 1023[5].

Currently, in agricultural applications, the use of SoCs are very rare. However, the advent of SoC has a lot of potential for the agriculture and farming domain. Firstly, the use of SoCs based sensor nodes instead of current day embedded multi-chip sensor nodes will increase the computation power, and decrease the energy consumption. Also, the size of the nodes will be less and thereby, portability of the overall system increases. Compared to multiple silicon dies in SiP, SoC is single die based, and thus, SoCs result in lesser size, but, higher cost[6]

In figure 2, it can be explained that on the placement of ports connected to each part of the system, this electrical design is made using Fritzing software.

- NodeMCU-ESP8266 to process moisture moisture sensor data (YL 69).
- Moisture soil moisture sensor (YL 69) when executed process will get the output of data will be sent to NodeMCU-ESP8266.
- On board for placement of ports connected to each component.
- Jumper cable to connect all components to be connected

IOP Conf. Series: Materials Science and Engineering **621** (2019) 012001 doi:10.1088/1757-899X/621/1/012001

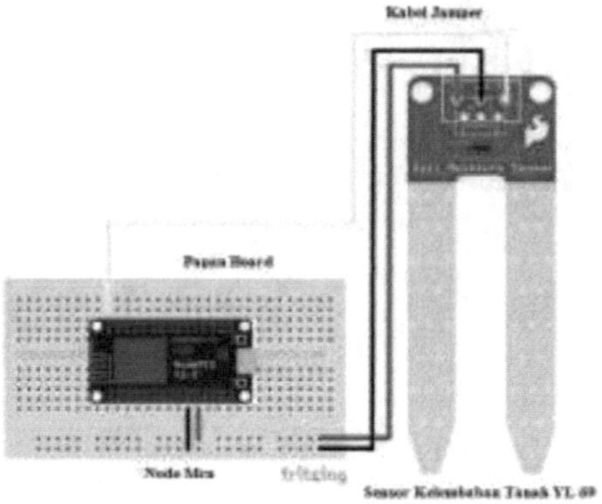

Figure 2. Arcitecture of Sensor Node

3.1.2. IoT platform

The IoT platform used is Node MCU esp8266 from Arduino. The microcontroller can be programmed using C/C++ programming language. The microcontroller has 12 general purpose input/output pin than can be used as input or output of the microcontroller. The microcontroller has 128KB memory and 4MB rom to store the program.

3.2. Design

3.2.1. Topology design

The network topology used in this research can be seen in Figure as follow:

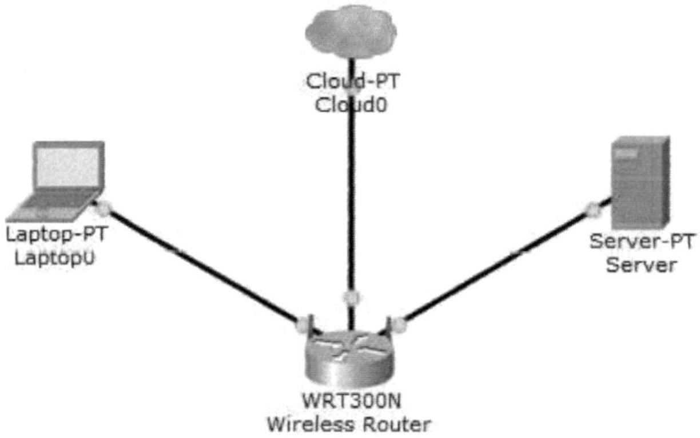

Figure 3. Topology design

The network topology used in this research are based on the client server topology. This topology uses one computer server, one wireless router, one laptop and an internet connection. This device was used so the system can be controlled remotely and the data can be viewed via web.

3.2.2. WSN design

The design of the WSN can be seen in Figure .

IOP Conf. Series: Materials Science and Engineering **621** (2019) 012001 doi:10.1088/1757-899X/621/1/012001

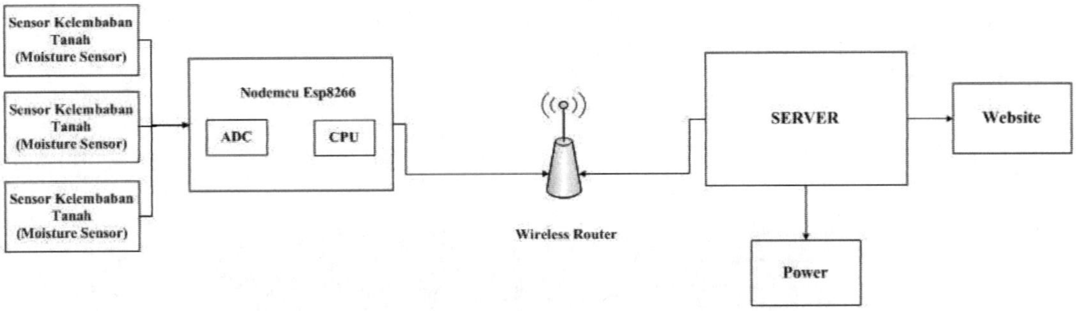

Figure 4. WSN design

Figure shows the design of the WSN. The sensors on the farm field sent its data trough a microcontroller to the server via a wireless connection. The data from the server can be seen remotely by accessing a website that provide the data from the farm field. The data from the server can be accessed through an internet connection.

3.2.3. Model design

In this research the overall WSN model created as minimum as possible to optimize the performance of the system. The model can be seen in Figure as follow:

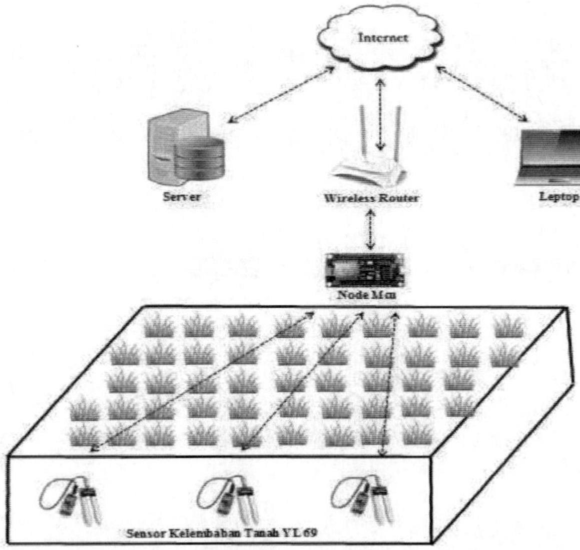

Figure 5. Model design

Figure shows the position of the sensor were spread in the model. The model of the farming field where using acrylic has 60 cm length, 90 cm wide and 20 cm high.

3.3. Simulation

The simulation of the system conducted by testing the client server network. The server received the data from the sensor trough wireless router and stored to the database. The router used as the communication media between the sensors and the server. The data stored from the moisture measurement process by the sensor can be seen through laptop.

3.4. Implementation

Shown the measurement of each sensors at the different condition. The condition set with normal, low moisture and high moisture. The data shows that the system able to read the moisture level of the farm

IOP Conf. Series: Materials Science and Engineering **621** (2019) 012001 doi:10.1088/1757-899X/621/1/012001

field successfully. The high reading of the sensor indicate that the soil has a high moisture and the low reading of the sensor indicate that the soil has a low moisture. The implementation of the model can be seen in Figure 6 below

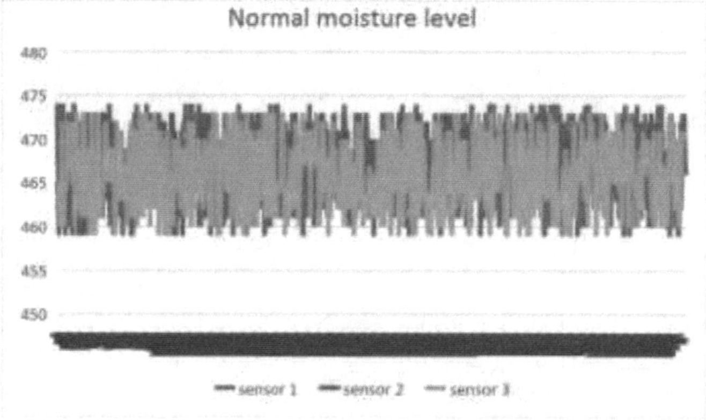

Figure 6a. Normal moisture level

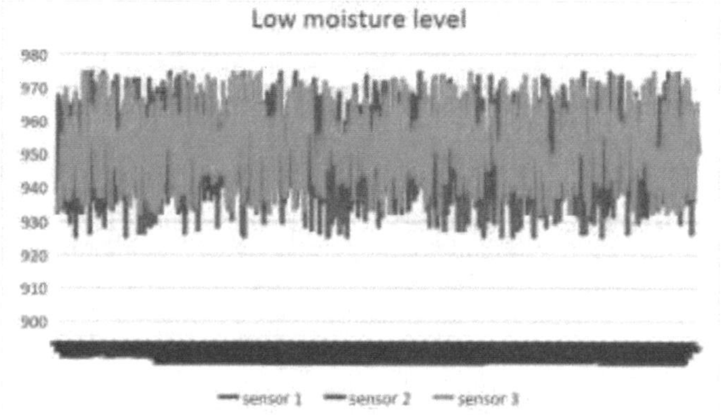

Figure 6b. Low moisture level

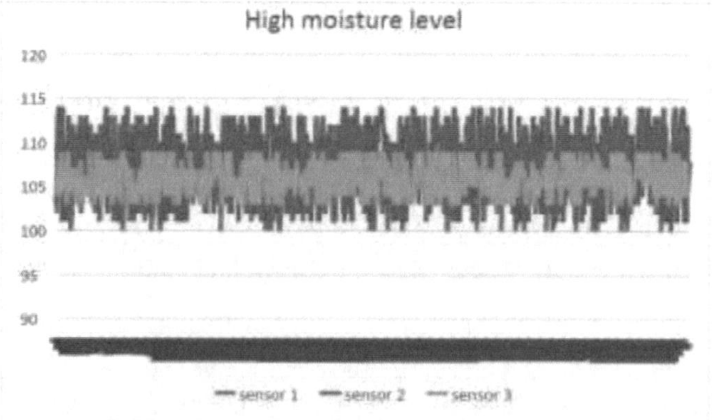

Figure 6c. High moisture level

Figure 6. Sensors reading

Figure 6 shows the measurement of each sensors at the different condition. The condition set with normal, low moisture and high moisture. The data acquisition process takes time every 5 minutes for two days. The data shows that the system able to read the moisture level of the farm field successfully. Figure 6a shows the reading at the normal moisture field, the data reading shows an average reading at 466.53 that indicate the soil is in normal moisture level. Figure 6b shows the reading at the low moisture field, the data reading shows an average reading at 951.35 that indicate the soil is in low moisture level. Figure 6c shows the reading at high moisture field, the data reading shows an average reading at 106.79 that indicate the soil is in high moisture level.

3.5. Monitoring

The monitoring phase is to monitor the connection among the IoT platform, the server and the client. The connectivity shown in Figure .

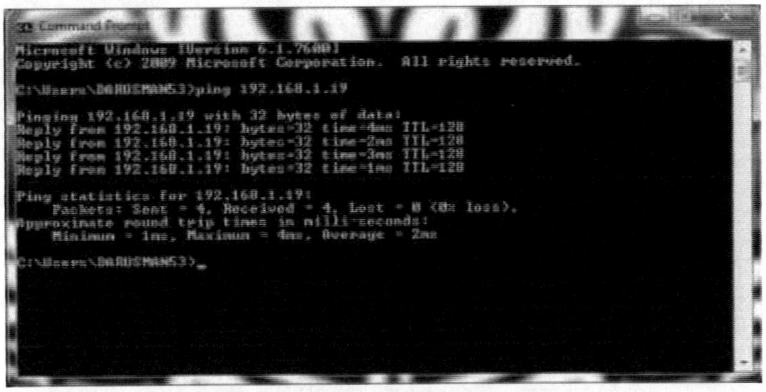

Figure 7. Connectivity monitoring

Figure shows the connectivity of the WSN system. The connection test was conducted by using a ping command to the IoT platform trough the server from the client PC. the monitoring process of the connection shows a stable connection and able to transmit data from the IoT platform to the client

3.5.1. Management

The configuration of the dynamic IP in the browser conducted to test the website user interface. The testing of the dynamic IP configuration shown in figure 6.

Figure 8. Client user interface

4. Conclusion

The WSN server model for smart farming optimization are using moisture sensors and an IoT platform to send the data from the sensors to the server. The sensor data can be access in the website trough an internet connection. The server role as a data storage to store the data from the moisture sensor that

IOP Conf. Series: Materials Science and Engineering **621** (2019) 012001 doi:10.1088/1757-899X/621/1/012001

spread across the farm field. the data in the server can be accessed remotely and provide a real time data from the farm field.

The sensor data from the farm field shows a promising reading. The sensor can acknowledge the moisture level of the soil in the farm field. the average normal moisture level of the soil is 466, the average high moisture level of the soil is 954 and the average low moisture level of the soil is 105.

References

[1] D. Ardiansyah 2017 "*Iot Framework For Smart Agriculture to Improve Agricultural*" Proceeding 12th Adri 2017 International Multidisciplinary Conference and Call For Paper.Pp.1-5. August.

[2] C. N. Cabaccan, F. R. G. Cruz, and I. C. Agulto 2017 "Wireless Sensor Network for Agricultural Environment Using Raspberry Pi Based Sensor Nodes," *2017ieee 9th Int. Conf. Humanoid, Nanotechnology, Inf. Technol. Commun. Control. Environ. Manag.*, Pp. 1–5

[3] C. Arif, B. I. Setiawan, And M. Mizoguchi 2014 "Penentuan Kelembaban Tanah Optimum Untuk Budidaya Padi Sawah Sri (System Of Rice Intensification) Menggunakan Algoritma Genetika Determining Optimal Soil Moisture for System Of Rice Intensification Paddy Field Using Genetic Algorithms Oleh :," Vol. 9, No. 1, Pp. 29–40

[4] I. John Wiley & Sons 2004 "The Network Development Life Cycle".

[5] S. M. Recorder 2012 "Soil Moisture Recorder," No. 603, Pp. 2011–2012

[6] T. Ojha, S. Misra, And N. S. Raghuwanshi 2015 "Wireless Sensor Networks For Agriculture: The State-Of-The-Art In Practice And Future Challenges" *Comput. Electron. Agric.*, Vol. 118, Pp. 66-84

IOP Conf. Series: Materials Science and Engineering **621** (2019) 012002 doi:10.1088/1757-899X/621/1/012002

Optimization of Metagenome Sequence Identification with Naive Bayes and Certainty Factor

Dian Kartika Utami[1], Herfina[2] and Iyan Mulyana[3]

[1] Diploma, Universitas Pakuan, Bogor, Indonesia
[2,3] Fakultas Matematika dan Ilmu Pengetahuan Alam, Universitas Pakuan, Bogor, Indonesia

E-mail: diankartikautami@unpak.ac.id

Abstract. Metagenome studies are an important step in taxonomic grouping. Taxonomic grouping can be done using the binning method. Binning is a process to determine the contigs of each group of phylogenetic species. In this study, Binning was carried out using the Supervise Learning approach. We use the Naïve Bayes Classifier method and Certainty Factor. The classification process is carried out on phylum taxon levels. many of the organisms used were 50 organisms and the length of the fragments used was 500 bp and many readings were 1000 readings. The accuracy results obtained by the Naive Bayes method are 62.5%. While the accuracy obtained in the Certainty Factor method is 54.45%. From the results of the two methods of testing, it can be concluded that Naive Bayes is the best method of classification compared to Certainty Factor.

Keywords: Metagenome Sequence, Binning, Naïve Bayes, Certainty Factor, Metagenome Sequence

1. Introduction

Metagenom is a microorganism which is based on the sequencing and analysis of the DNA of a community of microorganisms in nature, such as land, air, sea, or the entrails of living beings. The scope for the genome is based on the extent and depth of the environment of the genome [1]. The DNA sequencing process of the microorganism community directly produces fragments from various organisms that mix, so it is necessary to do a binning process to reduce errors in DNA assembly. Based on composition and homology the binning method consists of two approaches [2].

Based on research that has been done [3] to classify metagenom fragments using training data from 381 organisms. The method used is a Multiclass Support Vector Machine (SVM) with Spaced K-mers frequency as its features. Accuracy results at 400bp fragment length is quite good, namely 65.3% in genus taxon, 72.0% in taxon order, 78.2% in class taxon, 82.1% in phylum taxon. At 10Kbp fragment length the accuracy obtained is 95.4% - 97.6%. Studied the classification of metagenom fragments using the Naive Bayes Classifier (NBC) method and used 381 organisms and used variations in fragment lengths of 400 bp, 800 bp, 1 kbp, 3 kbp, 5 kbp and 10 kbp and used feature extraction mers. The fragment with a length of 400 bp resulted in an accuracy of 49.34% for extraction of 3-mer features and for extraction of 4-mer features to 53,95%, while for fragment lengths of 10 kbp the accuracy increased to 82.23% for 3-mer features extraction and for extraction of 4-mer feature to 85,89% [4]. The research conducted by Harun was a classification of metagenom fragments using the method Oblique Decision Tree method with genetic algorithm optimization (ODT-GA) [5]. The data used consists of only 10 organisms belonging to 3 genera. The features used are limited to $k = 2$, $k = 3$, and $k = 4$. For the length of the fragments used are 200 bp, 500 bp, 1 Kbp, 5 Kbp, and 10 Kbp.

Based on previous metagenom fragment classification studies, a study was conducted to optimize the identification of metagenomic sequences consisting of fragments of 50 organisms. Optimization using Naive Bayes and Certainty Factor.

2. Literature Study

2.1. Metagenome Sequence

Metagenomes are sequences and DNA analysis of microorganisms taken directly from the environment without a culture process. The fragments obtained from this metagenome contain various organisms so that the grouping process needs to be done to avoid errors in assembly. Taxonomic grouping is an important step in the metagenom studies [6].

2.2. Binning

Binning is one of the important steps for the assembly sequence in reaching contigs (adjacent fragments) which consist of a single grouping genome [7]. In binning there are 2 approaches, namely homology and composition. The homology approach is an approach that compares DNA fragments to databases which are references in aligning data sequences. Whereas composition-based approach has several advantages, namely as a shortcut to the need for sequence alignment, as input vectors resulting from extraction of base pair characteristics can be calculated as a composition feature and as input for observation by observation.

2.3. Naïve Bayes

Naïve Bayes is based on the Bayes Theorem, assuming that each feature in the classification does not depend on each other [8]. Naïve Bayes is used to calculate the probability of a class of each group of attributes that exist and determine where the most optimal class. The process of data classification with NBC illustrated in Figure 1.

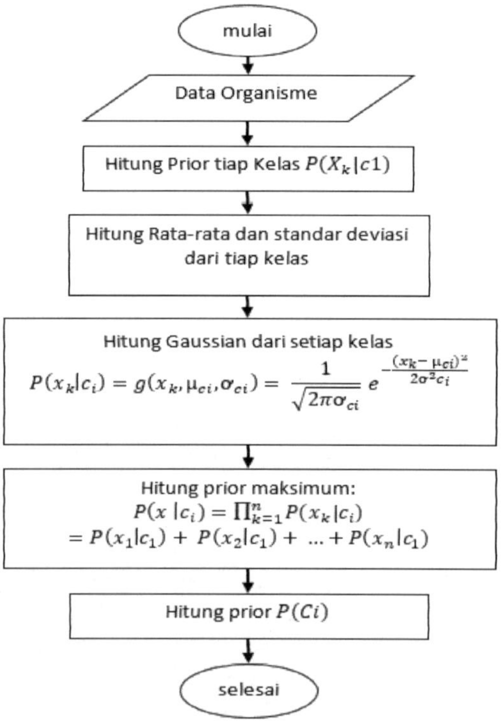

Figure 1. Classification stages with the Naive Bayes

2.4. Certainty Factor

Certainty Factor expresses confidence in an event (fact or hypothesis) based on evidence or expert judgment [9]. Certainty factor uses a value to assume an expert's degree of confidence in a data.

$$CF[H,E] = MB\,[H, E] - MD\,[H, E] \quad (1)$$

Information :

CF [H, E] = certainty of the hypothesis factor which is influenced by evidence e is known with certainty.

MB [H, E] = measure of believe on hypothesis H, if given evidence E (between 0 and 1).

MD [H, E] = measure of disbelieve against evidence H, if given evidence E (between 0 and 1).

The data classification process with Certainty Factor is illustrated in Figure 2.

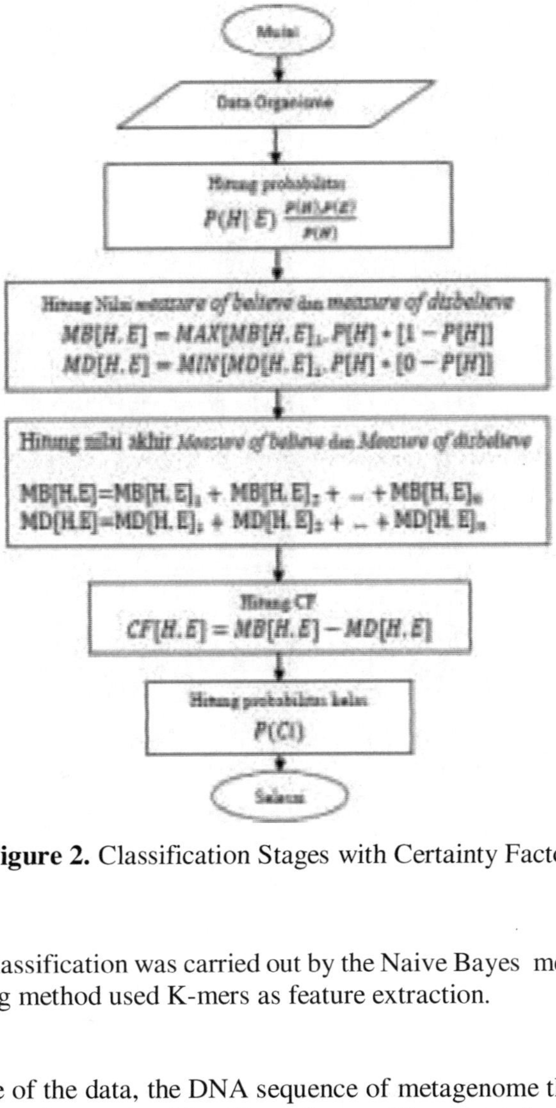

Figure 2. Classification Stages with Certainty Factor

3. Result

In this study metagenom classification was carried out by the Naive Bayes method and Certainty Factor, and the data pre-processing method used K-mers as feature extraction.

3.1. Pre-Processing

In the preprocessing phase of the data, the DNA sequence of metagenome that has been selected from the NCBI site will be broken down using MetaSim software. The steps taken using MetaSim are:

1. Fail containing the DNA sequences of microorganisms that have been downloaded from NCBI incorporated into the software MetaSim. The stages of inserting files into MetaSim software can be seen in Figure 3.
2. After entering the data from NCBI MetaSim into the software, the next process is to choose some of the DNA sequences of microorganisms that have been available in the database according to the needs of research. Database of DNA sequences of microorganisms can be seen in Figure 4.
3. The next step is to set the preset. The setting at preset done to set the length of fragments and the number of readings to be used. Preset settings can be seen in Figure 5.
4. The last step is to run the simulator according to the presets that have been set. The results of the DNA sequences of microorganisms that have been successfully described can be seen in Figure 6.

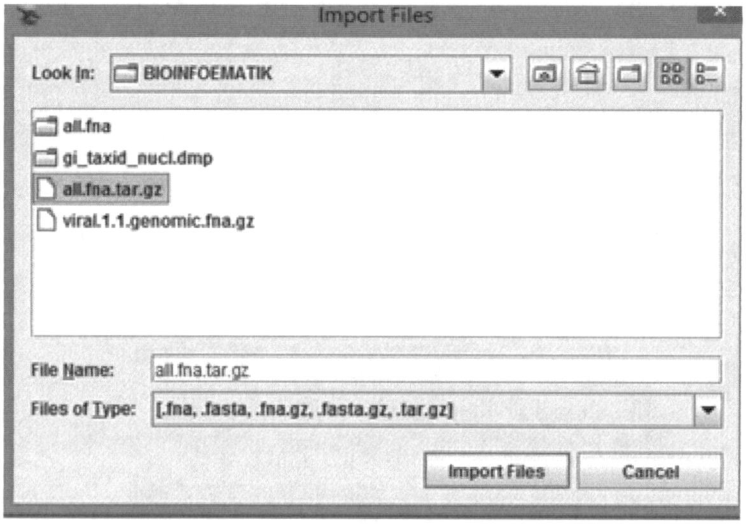

Figure 3. The process of entering metagenomic data on MetaSim

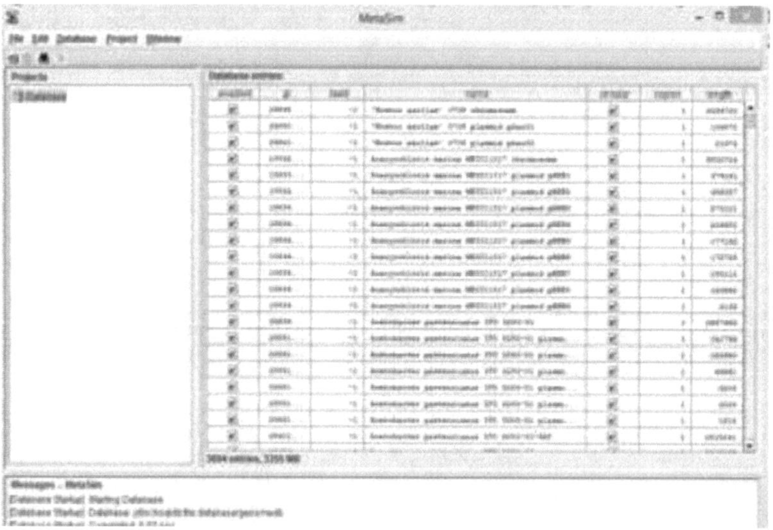

Figure 4. Database of DNA sequences of microorganisms on MetaSim

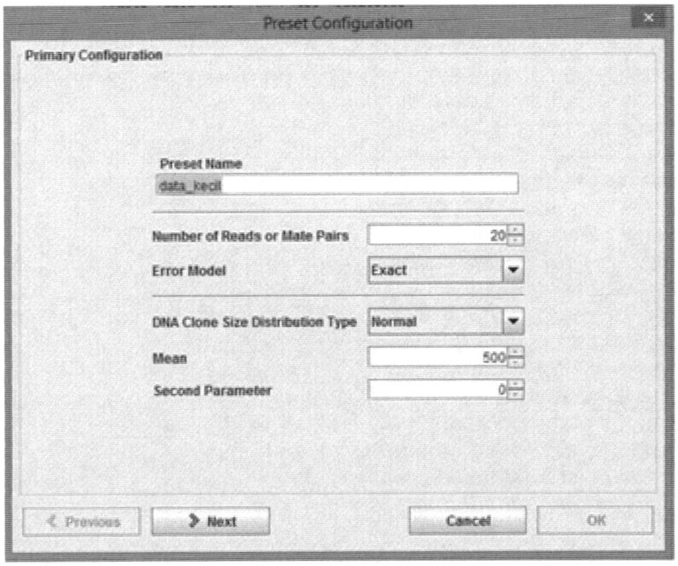

Figure 5. Preset

Figure 6. The results of the DNA sequences of microorganisms are successfully parsed

3.2 Data Fragment Metagenom

The processed data is obtained from the preprocessing data, the DNA sequences of metagenomes that have been selected from the NCBI site, and the fragments have been described using the MetaSim simulator. Data to be processed will be read many times according to research needs. In the example of this study, the prepared data was read 1000 times. This means that there are 1000 lines of microorganism fragments. The length of the fragment used is 500 base pair (bp). Fragment data that has been described using metasim can be seen in Figure 7.

IOP Conf. Series: Materials Science and Engineering **621** (2019) 012002 doi:10.1088/1757-899X/621/1/012002

Figure 7. Data Fragment yang sudah di uraikan menggunakan metasim

3.3 Feature Extraction

Feature extraction is done by reading the frequency of nucleotide combinations that may be formed by using k-mers for k = 4. The pattern of occurrence k is a pattern that displays k at a time in a sequence. The pattern of occurrence of k in the sequences is calculated using four main bases (A, C, T, and G) raised by the base pair sequence that you want to use (pattern of occurrence: 4 ^ k, with k ≥ 1). In this study k = 4 means that there will be 4 ^ 4 = 256 patterns of occurrence that are formed. An example for feature extraction using the 4-mers feature can be seen in Figure 8. And it illustrates the calculation results can be seen in Table 1.

3.4 Data Sharing

The data used in this study were divided into two parts, namely the training data and test data. The data used consisted of 50 organisms including 12 different phylum groups. Training data and test data are divided using k-fold cross validation with k = 4. The k-fold cross validation method repeats many times to divide a sample set randomly into mutually independent k-subset. Each replication has one subset used for testing while the rest is used for training.

From one thousand fragments, the division is divided into training data and the test data uses 4-fold cross validation. So there are 4 repetitions for each group of data and in each iteration 250 fragments become test data while the other 750 become training data.

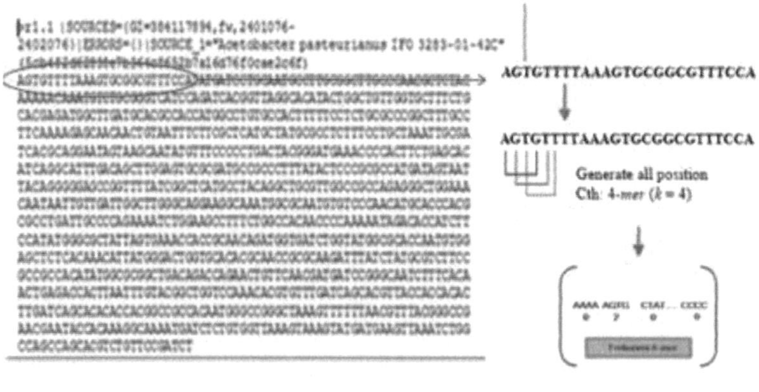

Figure 8. K-mers feature extraction

IOP Conf. Series: Materials Science and Engineering **621** (2019) 012002 doi:10.1088/1757-899X/621/1/012002

Tabel 1. Illustration of the results of k-mers frequency calculations on the data of 50 organisms that belong to the phylum 12 to 1000 readings.

No	ID	X1	X2	X3	X4	…	X256	Class
		AA AA	AA AT	AA AC	AA AG	… .	GGG G	
1	CCGCCTCGCGGCCAC CCGGTGTCAAGGGCG TGAGCGGCAGCGGCA ATCTCTGCG…………	2	0	0	2	… .	0	C1
2	ACGGATGGGAGAAGC GGCCAACGCAGTGGG GCGTGTTCATGGCGC ACTTCC……………… …	1	0	2	0	… .	3	C1
. . .	. . .	. . .	. . .	. . .	. . .	. . .	. . .	. . .
1000	AATTCCAAAATTTAA TTGGACATCATTGAG ATCTCTTAAAAAATT AAATA………………… …	17	17	2	3	… .	0	C12

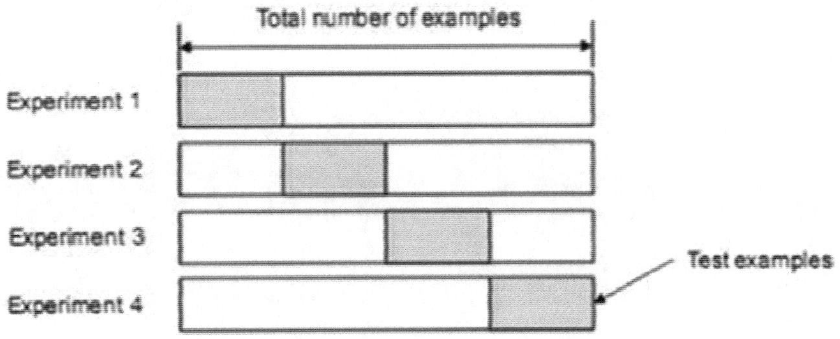

Figure 9. Illustration of k-fold validation [5]

3.5. Calculation of Naïve Bayes Classification

In the process of classification using Naive Bayes, Sequencing DNA data that has been extracted enters the first process. The first process in the classification Naive Bayes is looking for a probability value of each class by using the following formula:

$$P(x_k|Ci) = P(AAAA|C1) = \frac{\frac{2}{496582}}{\frac{1}{256}\cdot\frac{2}{3609} + \frac{1}{256}\cdot\frac{0}{2748} + \dots\dots + \frac{1}{256}\cdot\frac{0}{1685}} = 0.007532$$

$$P(x_k|Ci) = P(AAAT|C1) = \frac{\frac{0}{496582}}{\frac{1}{256}\cdot\frac{2}{3609} + \frac{1}{256}\cdot\frac{0}{2748} + \dots\dots + \frac{1}{256}\cdot\frac{0}{1685}} = 0$$

After the above process then looks up the value of the average and standard deviation. The following are the average values and standard deviations.

Calculate the average value of each class (μci):

$$\mu ci = \frac{x_1 + x_2 + x_3 + \cdots + x_n}{P(ci)}$$

$$\mu c1(AAAA) = \frac{0.007532 + 0.003766 + \dots + 0}{27} = 0{,}006117$$

$$\mu c1(AAAT) = \frac{0 + 0 + \dots + 0.002844}{27} = 0{,}002753$$

Calculate the standard deviation of each class (σci):

$$\sigma ci = \sqrt{\frac{\Sigma(X-\mu)^2}{N}}$$

$$\sigma c1(AAAA) = \sqrt{\frac{(0.001415)^2 + (-0.002351)^2 + \cdots \dots + \cdots ..}{27}} = 0{,}010687$$

$$\sigma c1(AAAT) = \sqrt{\frac{(-0{,}002753)^2 + (-0{,}002573)^2 + \cdots + \cdots}{11}} = 0{,}005285$$

Looking Gaussian value using the following formula:

$$P(x_k|c_i) = g(x_k, \mu_{ci}, \sigma_{ci}) = \frac{1}{\sqrt{2\pi\sigma_{ci}}} e^{-\frac{(x_k - \mu_{ci})^2}{2\sigma^2 c_i}}$$

Where:

$$\pi = 3{,}14$$

$$e = 2{,}71828$$

$$P(AAAA|C1) = g(x_k, \mu_{ci}, \sigma_{ci}) = \frac{1}{\sqrt{2(3{,}14 \times 0{,}010687)}} 2{,}71828^{-\frac{(2-0{,}006117)^2}{2\times 0{,}010687^2}} = 3{,}826334$$

After the input into the probability density function gauss then find the maximum value for each classification. Looking for maximum prior values using the following formula:

IOP Conf. Series: Materials Science and Engineering **621** (2019) 012002 doi:10.1088/1757-899X/621/1/012002

$$P(x \mid c_i) = \prod_{k=1}^{n} P(x_k \mid c_i)$$

$$P(x \mid C1) \quad = P(x_1|c_1) + P(x_2|c_1) + \ldots + P(x_n|c_1)$$

$$= P(AAAA|C1) + P(AAAT|C1) + \ldots\ldots + P(GGGG|C1)$$

$$= 3{,}826334 + 3{,}276811 + \ldots\ldots + \ldots\ldots = 35771{,}589737099$$

After determining the maximum value we determine the probability by dividing the maximum value Phylum the overall maximum amount. Next the results of the probability can be seen in Figure 10.

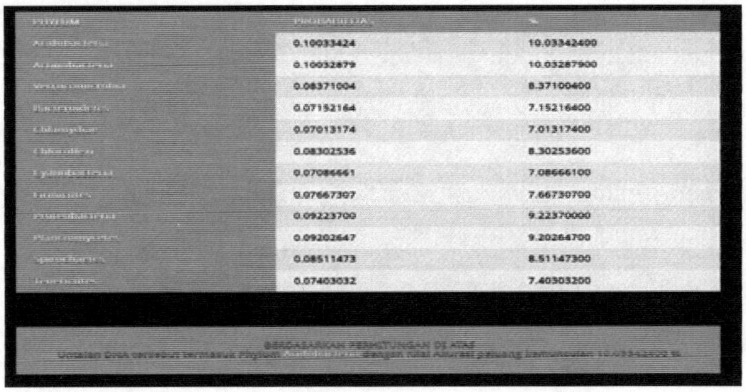

[illegible]	[illegible]	%
[illegible]	0.10033424	10.03342400
[illegible]	0.10032879	10.03287900
[illegible]	0.08371004	8.37100400
[illegible]	0.07152164	7.15216400
[illegible]	0.07013174	7.01317400
[illegible]	0.08302536	8.30253600
[illegible]	0.07086661	7.08666100
[illegible]	0.07667307	7.66730700
[illegible]	0.09223700	9.22370000
[illegible]	0.09202647	9.20264700
[illegible]	0.08511473	8.51147300
[illegible]	0.07403032	7.40303200

BERDASARKAN PERHITUNGAN DI ATAS

Figure 10. The results of the probability value

3.6. Calculation of the Certainty Factor Classification

In the process of classification using certainty factor, Sequencing DNA data that has been extracted enters the first process. The first process in the classification with the certainty factor is to find the probability values by using the following formula:

$$P(H|\ E)\ \frac{P(H),P(E)}{P(H)}$$

$$P(AAAA|C1) = \frac{\frac{2}{3609}}{\frac{1}{256}} = 0{,}141868$$

Value 2 (two) is an emergence pattern AAAA feature 2 (two) times, 3609 is the total value of the pattern emergence AAAA feature divided by 1/256 emergence pattern features base pairs. After the above process then looks up the value Measure of believe and disbelieve Measure of value. The following are the values of MB and MD.

Calculate the value of MB and MD by using the formula:

MB[H,E]=MAX[MB[H,E]_1,P[H]*[1-P[H]]
=[Max[0.141868 ,0.027]-0.027]*[1-0.027] = 0.111766

MD[H,E]=MIN[MB[H,E]_1,P[H]*[0-P[H]]
=[Min[0.141868,0.027]-0.027]*[0-0.027] = 0

The value of 0.027 is the average value of class C1. After the above process, then add the final value of Measure of believe and the value of Measure of disbelieve each data.

$$MB[H,E] = MB[H,E]_1 + MB[H,E]_2 + \dots + MB[H,E]_n$$

$$= MB(AAAA|C1) + MB(AAAT|C1) + \dots\dots + MB(GGGG|C1)$$

$$= 0.111766 + 0 + \dots\dots + \dots\dots = 54{,}506$$

$$MD[H,E] = MD[H,E]_1 + MD[H,E]_2 + \dots + MD[H,E]_n$$

$$= MD(AAAA|C1) + MD(AAAT|C1) + \dots\dots + MD(GGGG|C1)$$

$$= 0 + 0.000729 + \dots\dots + \dots\dots = 0.051179$$

Calculate CF by using the formula:

$$CF[H,E] = MB[H,E] - MD[H,E]$$

$$CF[H,E] = 54.50661 - 0.051179$$

$$= 54.454851$$

The value of CF 54.4548 is obtained from the reduction in the values of mb and md obtained from the previous stage. After obtaining the certainty factor value, the results are obtained along with the type of phylum.

3.7. Sensitivity and Specificity

In this study, in addition to the application of methods Naive Bayes and Certainty Factor in classifying sequences metagenom to seek maximum accuracy, it has also been conducted in the process of testing the accuracy of classification using Sensitivity and Specificity approach. Sensitivity is used to measure the proportion of true positives correctly identified, while specificity to measure the proportion of negatives which are correctly identified. Sensitivity and specificity calculations performed on each phylum to see the accuracy of the model was generated against each phylum. The results of calculation of sensitivity can be seen in Figure 11 and Figure 12, the calculation of specificity can be seen in Figure 13 dan 14.

IOP Conf. Series: Materials Science and Engineering **621** (2019) 012002 doi:10.1088/1757-899X/621/1/012002

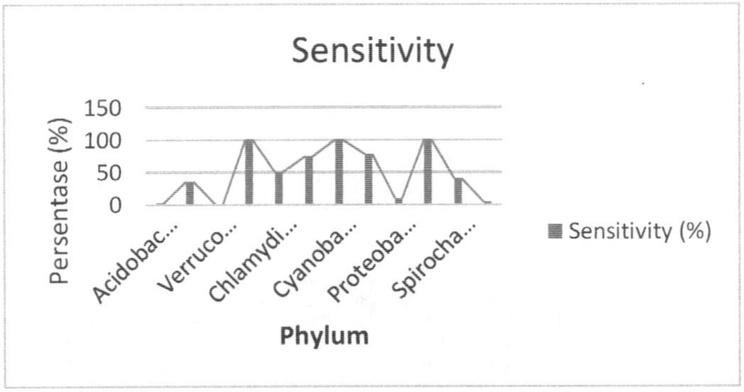

Figure 11. Sensitivity with Naïve Bayes Method

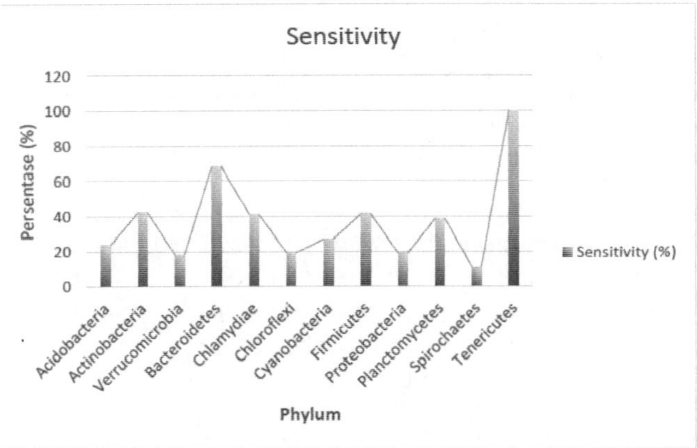

Figure 12. Sensitivity with Certainty Factor Method

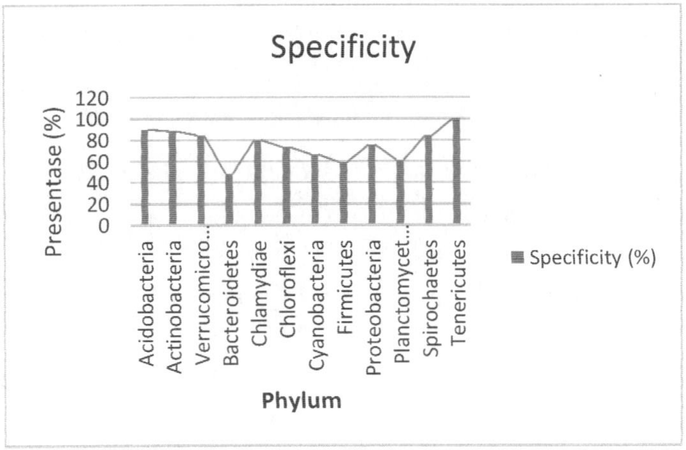

Figure 13. Specificity with Naïve Bayes Method

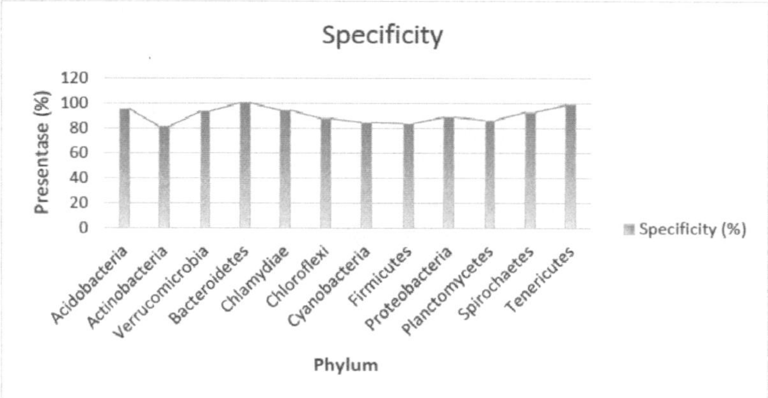

Figure 14. Specificity with Certainty Facor Method

4. Conclusion

The application of the Naïve Bayes method and Certainty Factor in the classification of metagenom uses 50 organisms belonging to the 12 phylum, namely: Acidobacteria, Actinobacteria, Verrucomicrobia, Bacteroidetes, Chlamydiae, Chlorophlexi, Cyanobacteria, Firmicutes, Proteobacteria, Planctomycetes, Spirochaetes, Tenericutes.

Based on the calculation of the values of sensitivity and specificity in the Naive Bayes method that have been carried out, the values obtained for sensitivity range from 0 to 100% while the values for specificity range from 48% to 100%. While based on the calculation of the value of sensitivity and specificity in the method of Certainty Factor, the value obtained for the sensitivity ranged from 11% to 100%, while the value for specificity ranged from 81% to 100%.

The application of the Naïve Bayes method to the classification of metagenome does not produce output that is in accordance with the class. Many sequencing DNA that is input is classified into another phylum. An example is the phylum acidobacteria the test data used by 27 sequences but only 6 sequences are read in the phylum correct. While the application of the classification method of Certainty Factor this metagenom DNA Sequencing widely read when in the classification of the phylum to which Bacteroidetes 202 sequences and a low of Verrucomicrobia read as much as 2 sequences in the phylum correct.

References

[1] Maureen A and O'Malley 2006 *Metagenomics*. Sydney-Australia. University of Sydney.

[2] Wooley J C, Godzik A and Friedberg I 2010 A primer on metagenomics. *PLos Computational Biology*. 6(2):1–13. doi: 10.1371/journal.pcbi.1000667.

[3] Ariny 2013 Klasifikasi Fragmen *Metagenome* Menggunakan Metode *Support Vector Machine* (SVM) [skripsi]. Bogor-Indonesia : Institut Pertanian Bogor.

[4] Utami D K 2014 Klasifikasi Metagenom dengan Metode *Naïve Bayes Classifier* [tesis]. Bogor-Indonesia : Institut Pertanian Bogor.

[5] Harun A S 2014 Kalsifikasi Fragmen *Metagenome* Menggunakan *Oblique Decision Tree* dengan Optimasi Algoritme Genetika [skripsi]. Bogor-Indonesia. Institut Pertanian Bogor.

[6] Higashi S, Barreto André da MS, Cantão ME and de Vasconcelos ATR 2012 *Analysis Of Composition-Based Metagenome Classification*. BMC Genomic 2012, 13(Supply 5):S1. http://www.biomedcentral.com/1471-2164/13/S5/S1

[7] Strous M, Kraft B, Bisdorf R and Tegetmeyer H 2012 *The binning of metagenomic contigs for microbial physiology of mixed cultures.* Frontiers in Microbiology 3.

[8] Han J, Kamber M 2001 *Data Mining Concepts and Techniques*. Cerra DD, Severson H, Breyer B, editor. San Diego (USA): Academic Pr. ISBN 978-0-12-381479-1.

[9] Turban E and Aronson J E 2001 Desecion Support System and Intelligent System, 6th. Edison; Prentice Hall International Edition, New Jersey.

IOP Conf. Series: Materials Science and Engineering **621** (2019) 012003 doi:10.1088/1757-899X/621/1/012003

The Effect of Chronic Energy on Increasing the Effectivity of Pregnant Women's Nutrition Using Decision Support Systems (DSS) with Analytical Hierarcy Processs (AHP)

Y Wahyuni[1,3], H Tus Sa'diah [2], C Zaddana [2]

[1]Department of Technology Computer, Faculty of Diploma III, Pakuan University, Jl. Ciheuleut Pakuan PO Box 452 Bogor 16143 West Java, Indonesia
[2]Department of Technology Computer, Faculty of Diploma III, Pakuan University, Jl. Ciheuleut Pakuan PO Box 452 Bogor 16143 West Java, Indonesia
[3]Department of Pharmacy, Faculty of MIPA, Pakuan University, Jl. Ciheuleut Pakuan PO Box 452 Bogor 16143 West Java, Indonesia

Corresponding author: yuli_wahyuni@unpak.id

Abstract. The Indonesian Ministry of Health in 1999 proclaimed LILA as a measure of the upper arm circumference used by pregnant women in childbearing age as a chronic energy deficiency indicator (SEZ), with a LILA limit of <23.5 the risk of SEZ. Chronic energy deficiency can cause low birth weight babies (LBW), so pregnant women will be at risk of experiencing difficulties and pain during the birth process. The lack of technology that can quickly determine how pregnant women experience Chronic Energy Deficiency has an impact on increasing the mortality rate in pregnant women and the conceived fetus. Some variables of effectiveness of nutritional intake of pregnant women such as diet, exercise, vitamins C, D, and E are the nutrients needed during pregnancy. There is no proper formulation to know the nutritional needs of pregnant women and nutritional quality factors are different, so nutrition experts desperately need a petrified system to analyze quickly, accurately, easily, effectively and efficiently in determining the best decisions from various alternatives for pregnant women who lack symptoms of energy Chronic (KEK) by combining analytical data and models on computerized systems using multicriteria Decision Support System (DSS) method, namely Analytical Hierarchy Process (AHP). This study designed the DSS using AHP method by combining Expert Choice in it. DSS research is designed not to make decisions, but as a tool that helps to support decision making.

1. Introduction

The Indonesian Ministry of Health in 1999 proclaimed LILA as the upper arm circumference measure used by pregnant women at childbearing age as a tool of Chronic Lack of Energy indicator (KEK), with LILA limit <23.5 SEZ risk [1]. During pregnancy is very dangerous a pregnant woman if chronic energy deficiency. Chronic energy deficiency can cause low birth weight babies (LBW), so that pregnant women will be at risk of difficulty and pain during childbirth. The lack of technology that can quickly detect pregnant women Chronic Less Energy has an impact on increasing mortality rates in pregnant women and the fetus is conceived. Research Sumarno et al also showed that pregnant women KEK in West Java have a risk of giving birth to LBW 2.3 times compared with pregnant women who KEK. Whereas LBW has a high risk of death seita will experience growth disorders (45) [2].

Some variables of the nutritional intake of pregnant women such as Diet, Sports, Vitamin C, D, and E are the nutrients needed during pregnancy. Nutritional needs of pregnant women in intake food required fetus during pregnancy if there is no proper formulation then it can lead to Chronic Energy Lack. Maternal nutrition supporting factors are basically the same, but the nutritional value level is different, so the quality or quality of nutrition is also different. Given the same variety of factors, but there are different levels of factor values in each pregnant woman, the determination of pregnant women with chronic energy deficiency symptoms becomes difficult, so nutrition experts desperately need a system that can help to analyze quickly, precisely, easy, effective and efficient in determining the best decision from various alternative pregnant women who suffer from symptoms of Chronic Energy Lack (KEK). In 2010 Yuli Wahyuni researchers have successfully found the right method used for decision support system that is Analytical Hierarchy Process (AHP) method which is effective and efficient to solve multikreterian problem in decision support decision system especially for pregnant mother nutrition with variable of calcium, energy, vitamin A, phosphorus, folic acid, vitamin K, iodine, iron and soup that must be consumed by mother hami (salmon soup, tomato soup, asparagus crab soup, white mushroom soup, red bean soup and pumpkin cream soup) [4] .

The development of research conducted by Yuli Wahyuni in 2017 has succeeded in making DSS application which helps nutrition experts in making decision by using basic nutritional data data, namely: dietary variable, pregnant women's exercise variable, vitamin C, vitamin D and vitamin E, while alternative soup choices include: alternative soups of choice for pregnant women: salmon soup, tomato soup, asparagus crab soup, white mushroom soup, red bean soup, yellow pumpkin soup accompanied by ingredients contained [5]. Sumarno in 2005 in his research revealed that Increased access to health services can improve the state of health and nutrition. In addition, efforts to improve the state of health and nutrition can take the form of treatment and prevention [3]. Examples of cases are categorized cases of multicriteria (compound criteria) / complex. The multicriteria decision-making method is Analytical Hierarchy Prosess (AHP), where factors influencing decision making are analyzed in the structure of tiered problems by assessing and ranking alternative decisions. The computerized system Decision Support System (DSS) combines data and analytical models for semiterstructured or unstructured issues. In this research the design of DSS using AHP motode, making the application using MS Access as DBMS database subsystem and while Microsoft Excel on the model subsystem and calculation process is integrated through VBA (Visual Basic For Application), to solve the problem, without covering the maintenance step (maintenance) and adaptation. In DSS research that is designed not to make decisions, but as a tool that helps to support decision-making.

2. Materials and Methods

2.1. Measured Variables

In this research the measurement of Chronic Energy Lose indicator (KEK) using height, upper arm circumference and high Rahim. While the variables of nutritional intake of pregnant women such as Calcium, Energy, Vitamin A, Phosphorus, Folic Acid, Vitamin K, Iodium, Iron, Diet, Sports, Vitamin C, D, and E.

2.2. Models Used

In this research, the model used is Analytical Hierarchy Process (AHP) which later as the basis of simulation in decision making.

2.3. Research Design

a) Designing AHP-based DSS system, this is tailored to the needs of system design that includes:
b) Planning Sub System Model Base.
c) Measurement factors of maternal nutrition selection along with their importance level measurement, problem solving with AHP method and creating enterprise model.
d) Planning Sub Data Base System.
e) Create a list of entities, create DFD (Data Flow Diagram), create ERD (Entity-Relationship Diagram) and database design.
f) Planning Sub User Interface System.
g) Menu and form design.
h) Planning Algorithm with Flow Chart.

IOP Conf. Series: Materials Science and Engineering **621** (2019) 012003 doi:10.1088/1757-899X/621/1/012003

2.4. Data Collection

In collecting this data, the data collected in the form of primary data consists of:

a) Data of pregnant women

b) Data factors that affect the nutrition of pregnant women

c) Data of nutrient intake in the form of variable data of Calcium, Energy, Vitamin A, Phosphorus, Folic Acid, Vitamin K, Iodium, Iron, Diet, Sport, Vitamin C, D, and E

d) The primary data collection techniques as follows: Interviews, literature and system testing

e) System test

Verification, validation and prototype.

3. Results

The application of the Nutrition Information system for pregnant women can improve maternal health services from nutritional factors, thereby reducing the mortality rate of pregnant women and Chronic Energy Deficiency.

A. Problems and Solutions

a) Solving Model Database Problems

1. Structure Preparation Hierarchy

In this AHP application, the goal to be achieved is to obtain the best results in deciding a pregnant woman Chronic Energy Deficiency, based on the factors manifested in three alternative options, namely height, upper arm circumference and high uterus Rahim.

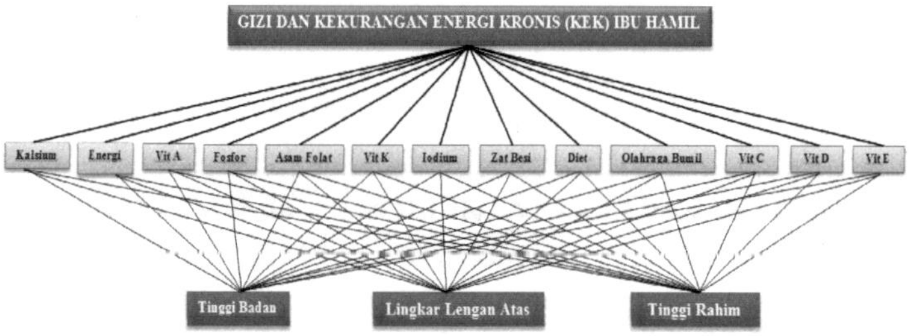

Figure 1. Hierarchy of Process Objectives

2. Factor Calculation Result

Here is a picture of the calculation of interests between factors of nutritional selection of pregnant women based on AHP scale with the result of 0.08 which means that the judgment data judgment has been correct. The exact calculation standard in determining the ratio value is less or equal to 0.1.

3. Result of Factor Calculation Graph

Here is a graph of the calculation of factors Selection of nutrition and chronic energy shortage of pregnant women.

IOP Conf. Series: Materials Science and Engineering **621** (2019) 012003 doi:10.1088/1757-899X/621/1/012003

Figure 2. Calculation of Factor Value

4. Selection of Pregnancy Selection of Pregnant Women

Here are some of the system application display of Nutrition Information for pregnant women:

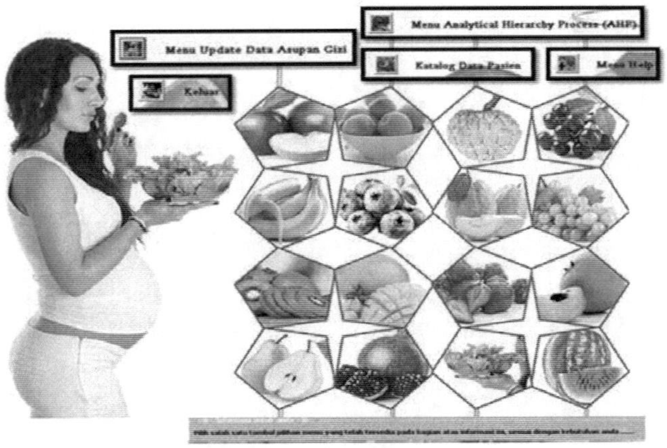

Figure 3. Display Main Menu

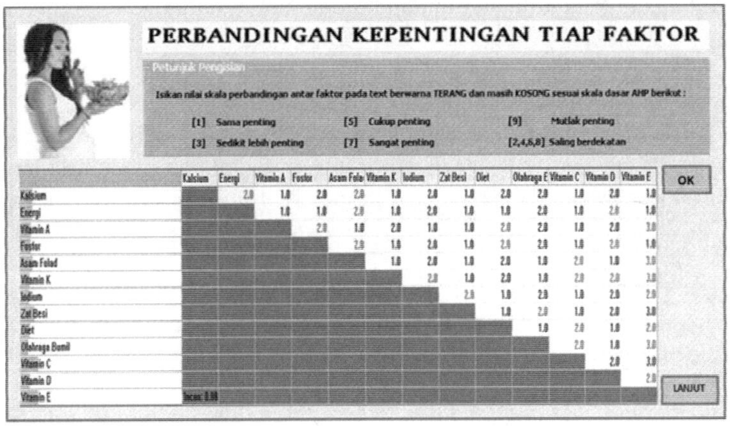

Figure 4. Input Number Value Scale Comparison Interest Factor Level

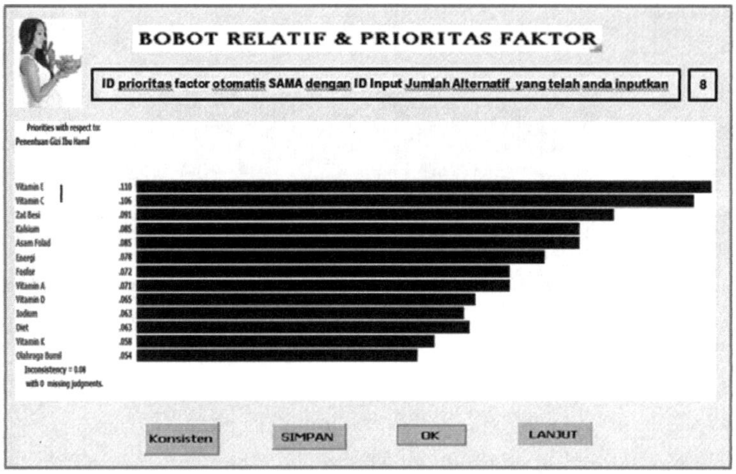

Figure 5. Relative Weight and Priority Value Factor along with the Graph

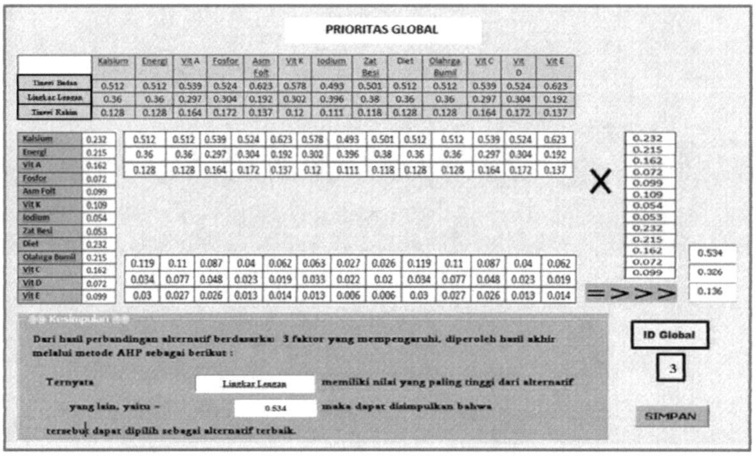

Figure 6. Global Priority Values and Alternative Conclusions

4. Conclusion

1. Application of Decision Support System has been successfully executed using AHP method with measurement of Chronic Energy Chronic indicator (KEK) using height, upper arm circumference and high Rahim. While the variables of nutritional intake of pregnant women such as Calcium, Energy, Vitamin A, Phosphorus, Folic Acid, Vitamin K, Iodium, Iron, Diet, Sports, Vitamin C, D, and E.

2. On the basis of existing data, built sub system:

   a. Databases consisting of patient data tables, complaint tables, login tables, alternative input tables, nutrition variable tables, factor priority tables, alternative priority tables, and global priority tables.

   b. A base model using AHP, by comparison of the importance level between factors, the determination of relative weight and priority for each alternative, the consistency calculation of the comparison of interests between factors, the comparison of the importance of all alternatives to each factor, the determination of the relative weights and the alternative priorities on each factor, and determination of overall rank (global priority).

c. Designing a sub-system of user interfaces, especially the visualization of the final result of global priority that shows the final conclusion that pregnant women know Chronic Energy Less in him so that it can prevent low birth weight babies (BBLR) early on. DSS design uses AHP method by combining Expert Choice in it. DSS research is designed not to make decisions, but as a tool that helps to support decision making.

3. System testing is done by:
   a. Verify with the result that the mathematical function for the calculation of the number of factor scale values has been properly realized in the design of the pre-designed model sub-system.
   b. Validation resulted in the conclusion that the application has been realized and able to produce the best solution and with this system nutrition experts in the community feel helpful in providing alternative solutions nutritional choices so as to avoid Chronic Energy Deficiency in Pregnant Women who can minimize low birth weight (LBW) from an early age.
   c. Testing prototype produces that the system has advantages in terms of ease, completeness, accuracy, and effectiveness.
4. System testing is carried out by:
   a. Verification with the results that the mathematical function for calculating the number of factor scale values has been realized correctly in the design of the model sub-system that has been previously designed.
   b. Validation generates the conclusion that the application has been realized and is able to produce the best solution and with this system nutrition experts in the community find it helpful in providing alternative nutritional solutions so that they can avoid Chronic Energy Deficiency in Pregnant Women who can minimize low birth weight babies (LBW) early.
   c. The prototype testing resulted that the system has advantages in terms of ease, completeness, accuracy, and effectiveness.

References

[1] Department of Health RI 1999 RI Guidelines. Guidance of the Upper Arm Circumference (LILA) in women of childbearing age. Jakarta: MOH RI.

[2] Sumarno, I, Sarasrmti E, Moses E, Yhazhazy, R Aryanti, et al. 1998 The risk of anemic pregnant women to deliver low birth weight babies (LBW). Bandung: Cooperation of Health Office Prov. West Java with Nutrition Research Center MOHRI.

[3] Sumarno 2005 Chronic Risk Factors of Chronic Energy In Pregnant Women In West Java (Advanced Analysis), PGM, Vol 28, No. 2, p. 66-73.

[4] Wahyuni Y, Santoso BP, Nurwasito H 2010 Decision Support System Development for Determining Nutrition of Pregnant Women Using AHP (Analytic Hierarchy Process) Method, J. Development of Informatics & Computer Management, Vol 1, No 2, 116-127 .

[5] Wahyuni Y, Ardita M 2017 Decision Support System System Improved Effectiveness of Nutritional Intake of Pregnant Women, Computing, Scientific Journal of Computer Science and Mathematics, Vol 14, No 1, Page 1-50, ISSN 1693-7554.

Optimisation of electronics and mechanics system of automatic egg incubator machine

J Iskandar[1], S Alrasyid[1], Eqi Nurhaqiqi[1], F Andria[2] and E T Tosida[3]

[1] Departement of Computer Technology, Pakuan University, Jl. Pakuan Bogor, West Java, Indonesia
[2] Departement of Management, Pakuan University, Jl. Pakuan Bogor, West Java, Indonesia
[3] Departement of Computer Science, Pakuan University, Jl. Pakuan Bogor, West Java, Indonesia

*Correspondence: johan_iskandar@unpak.ac.id

Abstract: In this research report result of optimisation of the electronic and mechanical system of an automatic egg incubator machine. The design of mechanical systems using two types of motors are stepper (one motor) and servo (3 motors) with the aim to produce a smooth movement of egg shelves in the maximum load that can be accommodated. In the electronic system used the thermostat as the controller of speed rate or maximum hatch temperature limit, while to maintain the humidity in the incubator moisture sensor used dht22 which integrated with a microcontroller. Also, the heat source on the incubator used two air heaters with 220 V / 65 W specifications. From the results of electronic and mechanics analysis per average the electrical power spent every 6 hours is 0.78 kWh. From the results of the driving test using a stepper and servo, the driver moves more optimally using a servo motor at a maximum load of 315 eggs (9 kg).
Keywords: hatching machine;electronic system;mechanical system;servo;a stepper motor.

1. Introduction

Egg hatching machine is one of the most widely used equipment in the field of poultry farms [1]. This machine is used to increase the quantity of egg hatching. Although many egg hatching machines are either manual, semiautomatic or automatic, the success rate or percentage of hatchability is low. One of the most influential causes is the uneven distribution of temperatures [2]. In this era of globalisation time and energy are so meaningful that use is so cared for to be as effective and efficient as possible. With the rapid development of technology, for the work of electronic equipment is increasing and encourage people to seek innovations in the provision of facilities and means to achieve these goals. One of them is the optimisation of electronic cage automatic hatch system to help the development of quail egg farmers [3]. This hatchery has one egg shelf that will hold 315 quail eggs. The advantages of this hatching machine are to accelerate the process of egg hatchery, effective and efficient in the process of quail egg dropper and efficient in the use of electric power [4][5].

Tia et al (2016) in the research that has been done successfully designing prototype of automatic atmega328 microcontroller based on DHT11 sensor that is by laying some eggs to be conditioned in the testing machine with the temperature in the incubator and checking for the first two hours with checking interval of 10 minutes with automatic playback technique with the help of servo

IOP Conf. Series: Materials Science and Engineering **621** (2019) 012004 doi:10.1088/1757-899X/621/1/012004

motor [6] [7] [8]. In this paper, the results of mechanical system optimisation with a comparison of the drive system with a servo motor and stepper and also in electronic systems have been analysed for the needs of the device to produce energy-efficient hatching machines..

2. Materials and Methods

In this study the tools and materials used are wood as the primary material, stepper motor load 10 kg, two servo motor each with load 10 kg, two 65 watt air heater, thermostat, Arduino Uno, dht22, LCD with i2c wiring, aluminium foil, wire.

The design of the hatching machine includes two stages of designing mechanical systems and electrical systems. In mechanical systems, the primary material used as housing is wood, wood is selected for its heat-resistant properties both within the incubator and from the environment. Next is on the drive system, selected two kinds of motors are servo and stepper because the rotation coordinates can be determined [9].

In the electronic system used two 65W heater at 220V voltage. Use of two air heaters to obtain a faster hatching temperature of 39-40 ^{0}C. Furthermore, in controlling the temperature so as not to exceed the maximum temperature of hatching used a thermostat device. In a humidity setting a dht22 moisture sensor is deprogrammed using an Arduino uno microcontroller. Fig. 1 shows how is the system works.

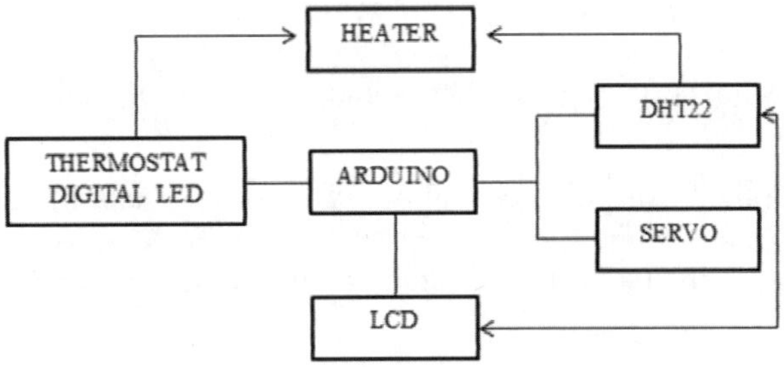

Figure 1. How the system works

3. Result and Discussion

3.1 Mechanical System

The Fig. 2 is the design of the hatching machine as well as Fig. 3 is the test of the heating laying in order to obtain a uniform heat distribution within the incubator.

Figure 2. Design of the hatching machine

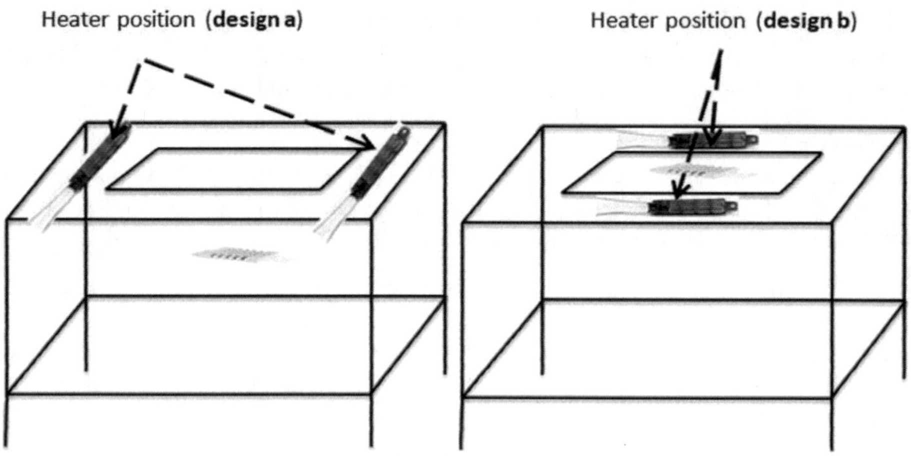

Fig. 3. Laying heater with the distance of bothsources is 35 cm (a) and 25 cm (b)

Testing of heating laying was done on several variations namely the distance between the heater and the distance between the air heaters with the shelf. In Fig. 3a the distance between the heaters is 35 cm, and the heating distance to the egg is 28 cm. At such distances, the heat distribution was evenly distributed at virtually every point of temperature testing with an analogue (mercury) thermometer with a difference of not more than 0.8 ^{0}C. While in Fig. 3b the temperature at the centre of the incubator always exceeds 40 ^{0}C when some test points have not reached 39 ^{0}C (optimum temperature hatching).

Servo servicing was done by varying the number of servo used and the servo laying position. The result is that using two servos placed at the centre point of each side of the shelf is more effective in moving the rack than using two servos placed on one side of the shelf.

3.2 Electronic System

In the design of electronic systems, things to consider is the power consumption of each component used in particular is in the form of heating. Initially before this research was conducted used four incandescent lamps with 220V / 100W specifications that can spend power of 9.6 kW every 6 hours. While this study reported using two 220V / 65W air heater can spend 0.78 kW every 6 hours of usage. However, one of the drawbacks of using the air heater is the time required to achieve 39 ^{0}C hatching temperature is relatively long in 60 minutes compared to using an incandescent lamp that is only 35 minutes.

In controlling and monitoring the temperature and humidity in the incubator initially using DHT22 sensor but obtained inaccurate in the measurement so its using thermostat for monitoring and to control the temperature while DHT22 still used to control the humidity. Completed electronics design is shown in Fig.4. The description of Figure 4 is as follows:

A. IC driver to controlling dc motor
B. Servos
C. Stepper Motor with 10 kg.cm of torque
D. Humidity sensor (DHT22)
E. Relay to controlling the air heater
F. Air heater
G. LCD to displaying the information of temperature, Relative humidity (RH)
H. AC power (220 V)
I. DC power (9 V) to supplying 3 of servos

IOP Conf. Series: Materials Science and Engineering **621** (2019) 012004 doi:10.1088/1757-899X/621/1/012004

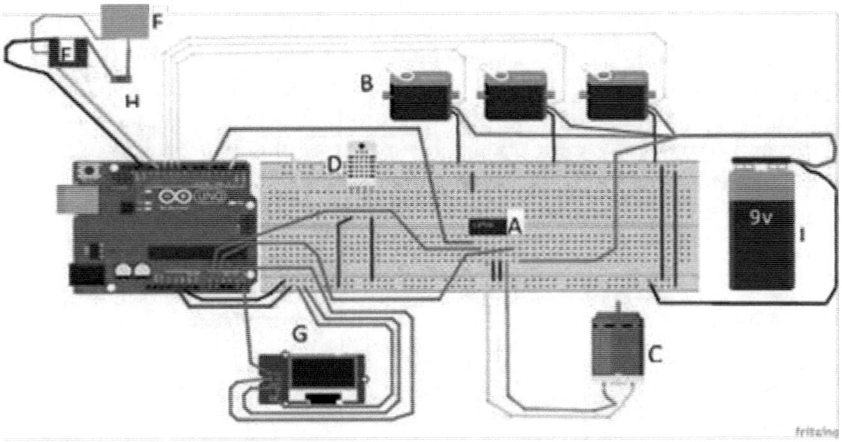

Figure 4. Electrical system design

Table 1 is the result of a servo simulation test using 6 hours and produced the first two movements right to 6 hours and the second to the left with 6 hours.

Table 1: Testing of motor movement

The first 6 Hour Test		**Second 6 hour Test**		**Third 6 hour Test**		**Fourth 6 hour Test**	
S	SM	S	SM	S	SM	S	SM
to right	to right	to left	to left	to right	to right	to left	to left

*SM = Stepper Motor, S = Servo

Table 2: DHT22 Simulation Test

Temperature above 39 ^{0}C	**Temperature below 37 ^{0}C**
Air heater was off	Air heater was on

Table 3: Testing of voltage variations on Motor

5 volt		**6 volt**		**9 volt**	
S	SM	S	SM	S	SM

The load that can be pulled was 8 kg	The load that can be pulled was 6.4 kg	The load that can be pulled was 9 kg	Can not move	Can not move (excess current)	Can not move

In the Table 2, a temperature sensor is tested. in this test the DHT22 temperature sensor detects changes in temperature in the incubator and the results are used as information to control the air heater. the test results show that the air heater will be automatically turned off when the temperature exceeds 39 ^{0}C and the heater is on again when the temperature is below 36 ^{0}C.

In the table above, we conducted an experiment with variations in electrical voltage by 5 volts, 6 volts, and 12 volts on the servo motor and stepper motor. Maximum load that can be driven by a servo or stepper motor at a voltage of 5 volts was 8 kg, then the 6 volt servo voltage can drive a 9 kg load while the stepper motor does not move at all, then when the voltage was increased to 12 volts both servo and stepper motors it does not move at all this is presumably because the current on the two motors is too large to exceed the maximum load current capable of holding.

4. Conclusion
From the results of the electronics and mechanics analysis performed the average electric power spent every 6 hours is 0.78 kWh. From the results of the driving test using a stepper and servo motors generated more optimal movement by using servo motor at the maximum load of 315 eggs (9 kg).

References
[1] Moreki JC and Mothei KM 2013 *Int. J. of Innovative Res. in Sci.* 2 5480
[2] C Ekwuozor, S & Sunday, Aguh, C Nzekwe, H 2008 *Nat. and Appl. Sci. J* 9 231
[3] Umar AB, Lawal K, Mukhtar M, MS Adamu 2016 *Int. J. of Modern Eng. Sci.* 5 11
[4] Benjamin N and Oye ND 2012 *Int. J. of App. Or Innov. In Eng. And Manag.* 4 90
[5] Frimpong K and Forson P 2017 *Int. J. of Comp. App.* 168 16
[6] Pravin K, Puja J, Poonam A, Pradnya P 2018 *Int. J. of Innov. In Eng. Res. And Technol, 2nd Conf. on Modern Trends in Elec. Eng, 30 – 31st March 2018* p 4
[7] Gbabo A, Liberty JT, Gunre ON, Owa GJ 2014 *Int. J. of Res. In Eng and Technol.* 3 522
[8] Fasanmi, Olubunmi 2013 *Int. J. of Scientific and Eng. Res.* 4 916
[9] Othman RA , Amin MR, Rahman S 2014 *J of Agri. And Veter. Sci.* 7 104

IOP Conf. Series: Materials Science and Engineering **621** (2019) 012005 doi:10.1088/1757-899X/621/1/012005

Forecasting Student Graduation With Classification And Regression Tree (CART) Algorithm

A Maesya[1] and T Hendiyanti[1]

[1]Faculty of Mathematics and Natural Sciences, Department Computer Science , Pakuan University
Jalan Pakuan PO.BOX. 452, Ciheuleut Bogor 16143

e-mail : a.maesya@unpak.ac.id , tissa.hendiyanti@gmail.com

Abstract. Forecasting Student Graduation with *Classification And Regression Tree* (CART) Algorithm aims to predict students' graduation in Computer Science Departement of Pakuan University. The method used is the CART Algorithm where this method works by forming a decision tree that can infer the rules of a particular classification of the decision tree. The results of this study resulted ouput in the form of decision trees and determine the students with predetermined criteria are classified in categories such as pass on time or pass not on time. Assessment of the accuracy of the data classification that has been done using confusion matrix validation technique. Once calculated with the matrix, the accuracy rate reached 77.5% in the training data.

1. Introduction

The development of technology and information has a very important role to support all work activities to be more optimal. Application of information technology can be used in data processing such as prediction or forecasting. In the process of processing data there are stages such as cleaning, selection, combination, until the data processed can be presented as a new information, this is called knowledge discovery in database (KDD). KDD is a series of processes to gain knowledge or patterns from data sets known as data mining [2]. Process of data mining include classification, association rule, and clustering. In the classification technique in describing the learning to do prediction on a value of the target in the category variable, one way in the classification that is often used is the decision tree. One of the methods contained in the decision tree is Classification And Regression Tree (CART) algorithm. This method works by forming a decision tree that can infer the rules of a particular classification that has a way of making decisions with regression activities on the data. Relevant studies related to predictions have been made by previous researchers, Mardiani [3] conducted the application of classification with the CART algorithm for predicting lectures for new students (Study Cased : STMIK MDP Palembang). This study uses data classification classification functionality to predict which students will continue to study and which do not continue in the next semester. Based on these studies formed a new study to forecasting students' graduation in Department Computer Science, Pakuan University.

2. Methods

The research method applied to Forecasting Student Graduation With Classification And Regression Tree (CART) Algorithm is the method of KDD (Knowledge Discovery In Database).

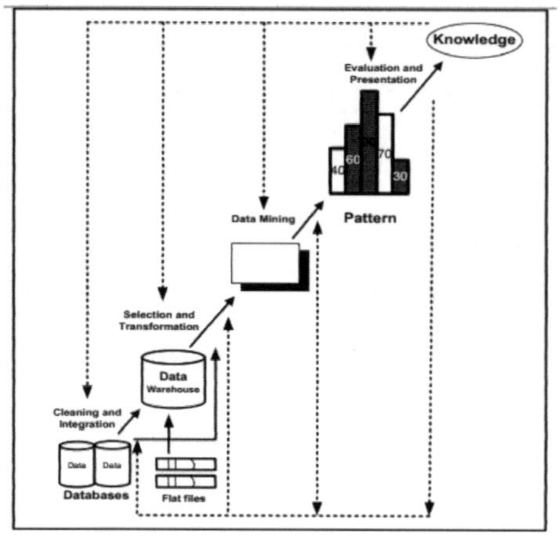

Figure 1. KDD Approach [1]

2.1 Cleaning and Integration Data phase

This process is an action to remove noise and inconsistent data or irrelevant data. The data set of students can be from PUTIK (Center of Information and Communication Technology) based on student data in SIMAK (Information System Academic) Pakuan University.

2.2 Selection and Transformation Data Phase

Data available on the database is often not all can be used, therefore the data to be retrieved is only relevant and relevant data. Later this relevant data will be the task of analysis to be returned into the database. Data is altered or merged into the appropriate format for processing in data mining.

2.3 Data Mining Phase

In this phase, different types of modeling techniques are selected and applied and the parameters calibrated to obtain optimal results. There are usually several techniques for the same type of data mining problem. Some techniques also have a need for a specific data form. Therefore, often the process of data preparation is needed again. The mining process is done by classifying the data with CART algorithm.

2.4 Pattern Evaluation Phase

The process of identifying patterns that represent knowledge based on multiple measures. The evaluation activities of the research measure the accuracy of CART algorithm using coincidence (confusion) matrix.

2.5 Knowledge Presentation Phase

Visualization and representation techniques a knowledge or information to the user. The last stage of the data mining process is how to formulate the decisions or actions of the analysis results obtained. Sometimes this should involve people who do not understand data mining. Therefore the presentation of data mining results in the form of knowledge that can be understood by everyone is a necessary step in the process of data mining.

3. Results

The results obtained from this research is the making of classification system of students' graduation forecasting of Departement of Computer Science, Faculty of Mathematics and Natural Sciences of Pakuan University using CART algorithm function in machine learning of SPSS Modeler application. The data is processed so that it will produce classification in the form of decision making tree when the student graduation is timely or not. The design of this system using Visual Studio 2015 application with the management of a. NET framework and database system using XAMPP (MySQL) application. Because it is not necessary to process data after predicting, this system can be used by anyone to obtain information on predictions of students graduation.

3.1 Page Input Test Data

The student data entry form will be classified using CART algorithm, resulting in the output in the form of the graduation level of the student pass, ie graduated on time (TW) or not on time (TTW). In the input menu test data can be directly viewed data entered in the table below the form fields. shown in Figure 2.

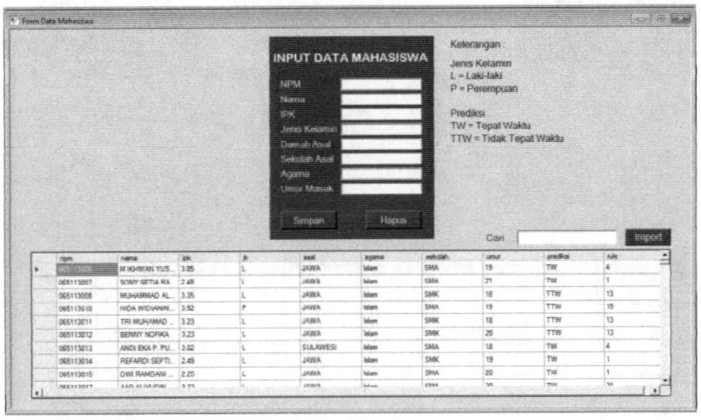

Figure 2. Display Page Input Test Data

In the input view of this test data, the user can retrieve data with the format .xlsx (excel worksheet) by clicking the "Import" button. The display of import menu is shown in Figure 3.

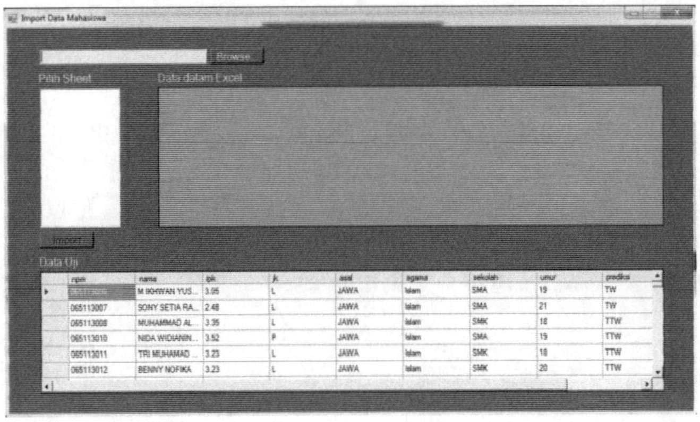

Figure 3. Display Page Import Test Data

3.2 Results of Classification Tree

The view of the classification tree provides an overview of the classification results of the previous training data. From the amount of data of 200 students, yielding decision tree and shown in Figure 4.

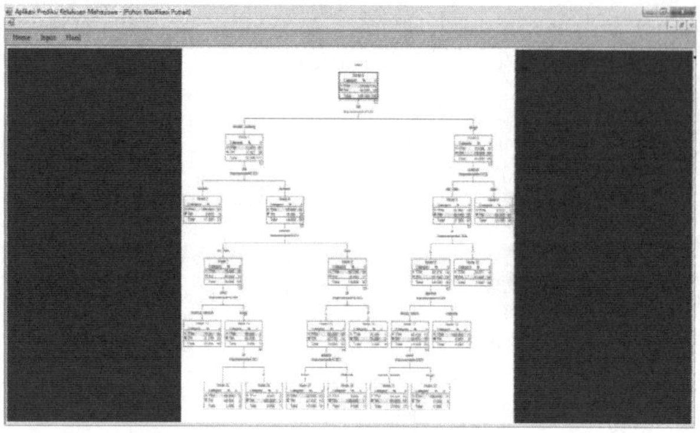

Figure 4. Display Results of Classification Tree

3.3 Result of Rule

The rules generated from the classification of training data 26 rules generated is shown in Figure 5.

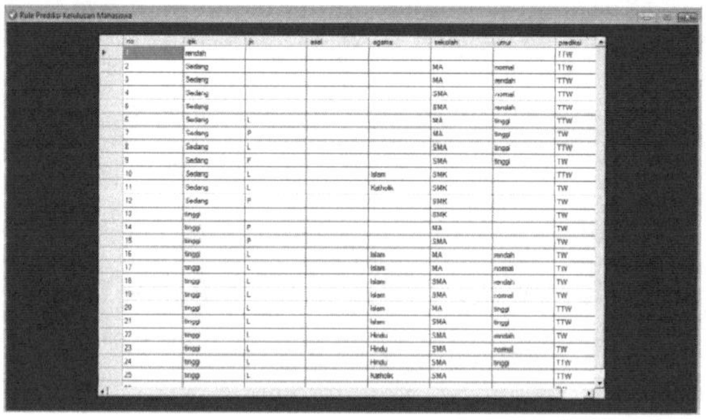

Figure 5. Display Result Rule of Test Data

3.4 Result Forecasting

The results of activities forecasting students' passing of the training data and test data are presented in Figure 6.

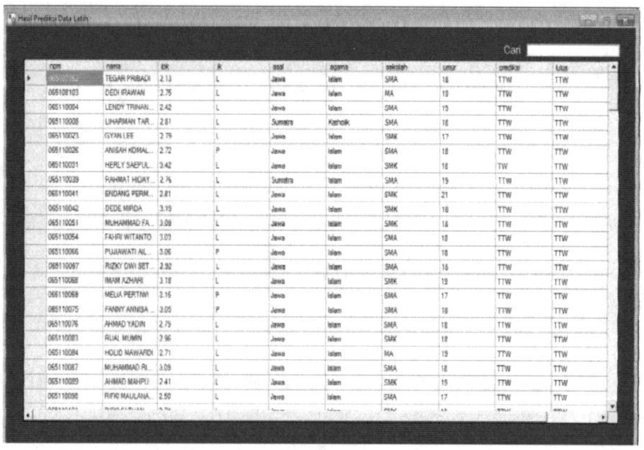

Figure 6. Display Result Forecasting

3.5 Test Validation

After the training data is processed, the analysis activity is done so that the results obtained in accordance with the desired design, while the analysis on this menu consists of analysis of

IOP Conf. Series: Materials Science and Engineering **621** (2019) 012005 doi:10.1088/1757-899X/621/1/012005

coincidence matrix (confusion), accuracy and prediction of test data. Each analysis result is presented in Figures 7, 8, and 9.

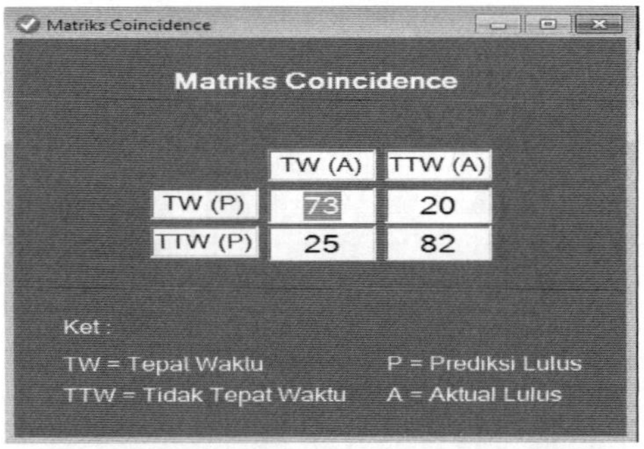

Figure 7. Display *Coincidence (Confusion) Matrix*

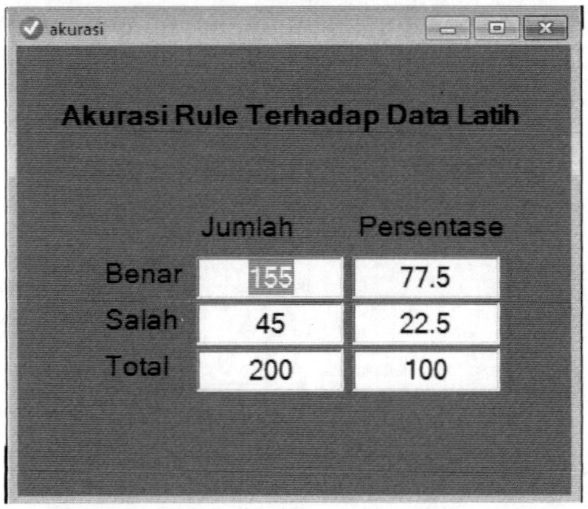

Figure 8. Display *Accuracy of Train Data*

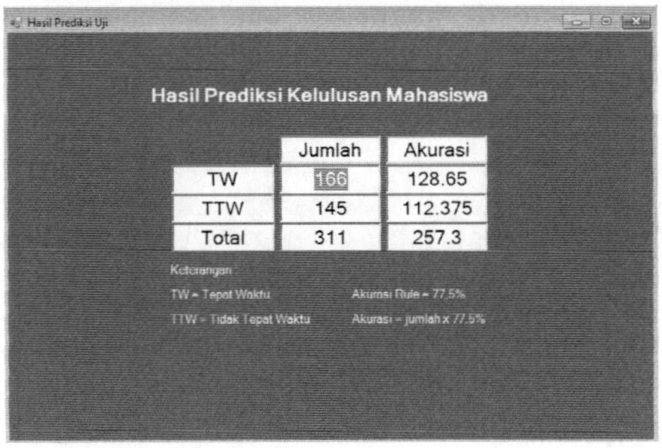

Figure 8. Display Result of Test Data

4. Conclusion

Forecasting Student Graduation with Classification And Regression Tree (CART) Algorithm aims to predict students' graduation in Computer Science Departement of Pakuan University. The method used is the CART Algorithm where this method works by forming a decision tree that can infer the rules of a particular classification of the decision tree. The results of this study resulted ouput in the form of decision trees and determine the students with predetermined criteria are classified in categories such as pass on time or pass not on time. Assessment of the accuracy of the data classification that has been done using confusion matrix validation technique. Once calculated with the matrix, the accuracyrate reached 77.5% in the training data.

IOP Conf. Series: Materials Science and Engineering **621** (2019) 012005 doi:10.1088/1757-899X/621/1/012005

References

[1] Han J and Kamber M 2012 *Data Mining Concepts and Techniques 2nd edition*. Mofgan Kaufhann Publishers, San Fransisco.

[2] Witten, Ian H, Frank, Eibe and Hal MA 2011 Data Mining: Pratical Machine Learning Tools and Techniques, Third Edition. Burlington: Morgan Kaufmann Publishers.

[3] Mardiani 2012 Application of Classification With CART Algorithm For Lecture Prediction For New Students (Case Study: STMIK MDP Palembang). Palembang. Thesis.

IOP Conf. Series: Materials Science and Engineering **621** (2019) 012006 doi:10.1088/1757-899X/621/1/012006

Air Pollution Monitoring for Bogor Smart City Based Internet of Things and Social Media (Twitter)

Mohamad Iqbal Suriansyah[1], **Prihastuti Harsani**[2]

[1] Department of Computer Science. Faculty of Natural Science and Mathematics, Pakuan University, Indonesia

[2] Department of Computer Science. Faculty of Natural Science and Mathematics, Pakuan University, Indonesia

Abstract. The increasing number of vehicles has affected the air quality of Bogor city, one of them is CO gas (Carbon monoxide). CO is included in air pollution variables when the levels are higher than 100 - 300 ppm [1]. In line with one of smart city development strategy City of Bogor is Smart Living to realize a decent, comfortable, and efficient living environment. Bogor City Government conducted traffic engineering around Bogor Botanical Gardens and Bogor Palace with the introduction of One Direction System (SSA) around the clock [4]. This has an impact on the concentration of air pollution around the SSA. The development of sensor and internet technology makes it easy to develop integrated pollution monitoring systems [3]. An Arduino-based air pollution monitoring system is used to detect CO gas. If the concentration of exposure received by the MQ-7 sensor module exceeds the prescribed normal limit the CO gas concentration will be transmitted over the internet and aired on a text-based social media twitter. The device is designed using some components consisting of Arduino, Ethernet Shield, and MQ-7 sensor module. Programs are written and uploaded using C programming languages and the Arduino IDE 1.0.6 compiler. This prototype has been tested in the laboratory and applied to measure CO gas in five subdistricts of Bogor City-road. The results of this system are expected to be used by policymakers as basic information for decision making to reduce pollution level of Bogor City.

Keywords : Air Pollution Monitoring, Bogor Smart City, Internet of Things, Social Media (Twitter)

1. Introduction

Along with the increasing physical development of the city, and motorized vehicles, air quality has changed. This is due to excessive air pollution. The number of motorized vehicles in the city of Bogor has continued to increase over time. In 2010 the number of motorized vehicles in the region was 270,845 units. In 2012, the number of motorized vehicles increased to 310,184 units. In a period of two years, the growth of motor vehicles increased by 12.7 percent [6].

An increase in the number of motorized vehicles is judged to be proportional to the increase in pollutants in the air. The pollution is one of undesired effect for human's health [11] since it has produced by vehicles [2] contains chemical residue of numerous pollutants [15]. Exhaust emissions such as NOx, SO2, CO2, and especially CO (Carbon monoxide) and others which could harm healthy life such as eye irritation, agitated, lung infection [7] and heart disease [13]. Motorized vehicles are the main source of CO pollutants which is around 59.2% [17]. This condition encourages the research for monitoring technology of air pollution in order to estimate the presence or absence of CO gas levels in the Bogor area.

Seeing the above problems, the design of a CO gas level monitoring tool is made that can provide information. This tool works based on the principle of local time and features, when detecting CO gas it will be processed by Arduino Uno and displayed through social media Twitter.

Previous research is monitoring of Carbon dioxide (CO2) in a room model based on the Atmega 8535 [1]. Input using TGS 4161 sensor to measure the absorption of CO2 gas [8], and the LM35 temperature sensor to measure temperature. The results of the design are then displayed on the LCD (Liquid Crystal Display) so that the user can read the pollutant measurement results. In this research a tool designed using Arduino Uno as a data processor, MQ-7 sensor to measure carbon monoxide gas, and an Ethernet shield as data communication to twitter. System testing was carried out on the highway which consisted of five districts in the city of Bogor which focused on measuring CO gas levels, the output was then arranged in the form of Twitter tweets.

2. Background

2.1. Air Pollution

Pollution is the entry of substances, energy, or other components in the environment that cause changes in the environmental order by human movements. Pollution means the entry of pollutants as a result ofhuman activities or natural rules found in place and improper amounts. Pollution can be found, for a case on land, water, air, even noise from motors, engines, and others. Air pollution is the presence of one or more physical, chemical or biological substances in the atmosphere in an amount that can endanger human [12], animal and plant health, disrupt aesthetics and encouragement, or damage property. Pence, air circulation can be caused by natural sources or human activities.

2.2. Carbon Monoxide (CO)

CO is an odorless and natural compound, which is combined into Carbon and Oxygen. This mixture is the result of incomplete combustion. often occurs in the combustion process in the engine. Carbon monoxide is formed when there is a lack of oxygen in the combustion process. Carbon monoxide can be bound to blood hemoglobin more strongly than oxygen which can form carboxyhemoglobin (COHb), which causes the supply of oxygen to the body's tissues to be blocked along with a range of ppm (parts per million). A few of people is not known about the relationship between exceedances of air quality standards for carbon monoxide and respiratory health disorders [18]. Table 1 explains the relationship between pollution levels and air quality

Table 1. Level of Carbon Monoxide [6]

Category	Range (ppm)	Carbon monoxida
Good	0-50	No Effect
Medium	51-100	Blood chemistry growth but not detected
Unhealthy	101-199	Increase in cardio vascular in smokers who have heart disease
Very Unhealthy	200-299	An increase in cardio vascular in non-smokers seems a real weakness
Harmful	More than 300	Very dangerous levels

2.3. MQ-7

MQ-7 is a gas sensor that is quite easy to use, this sensor has great sensitivity and gas selectivity. This sensor has high sensitivity and fast response time. The sensor output is analog resistance. MQ-7 can detect CO-gas absorptions everywhere from 20 to 2000ppm [16]. The output in the form of analog stability is required only to supply 5V power to the heater coil, add load resistance (RL), and connect the output to the ADC. MQ-7 gas sensor structure and arrangement is the sensor material, tin dioxide (SnO2). MQ-7 has 6 pins, 4 pins are used to pick up the signal, and 2 pins are used to provide heating of the sensor material. [9]

2.4. Arduino Uno

Arduino UNO is a microcontroller board based on ATmega328. Arduino UNO has 14 digital input/output pins (6 of them as PWM outputs) [10], 6 analog inputs of a 16MHz Crystal oscillator. a USB connection, a power jack, an ICSP header, and a reset button. Arduino UNO contains everything needed to support the microcontroller, easily connect it to a computer with a USB cable or supply it with an AC adapter to DC or use a battery to start it. [14]

2.5. Twitter

Twitter is a website that offers microblog social networking, allowing users to send and read messages. The word twitter actually means 'tweet'. Microblogging is for short update about something (whether it's an ongoing event or an update news) that is presented in a short character. Twitter is appear as a medium of choice for scientists research [5].

3. Method

This research was conducted from February 2016 to April 2016. The research was conducted in the city of Bogor. Research Material : *Arduino Uno R3, Modul MQ-7, Ethernet Shield*

3.1. Flow System

This system uses Arduino UNO that communicates with serial communication via an Ethernet shield. The system input is an MQ-7 sensor module that functions as a component to detect carbon monoxide gas concentrations. The general description of this system consists of 3 aspects, namely the system input using the MQ-7 sensor module. Control the system using Arduino Uno. The system output is the display of monoxide gas concentration on Twitter.

IOP Conf. Series: Materials Science and Engineering **621** (2019) 012006 doi:10.1088/1757-899X/621/1/012006

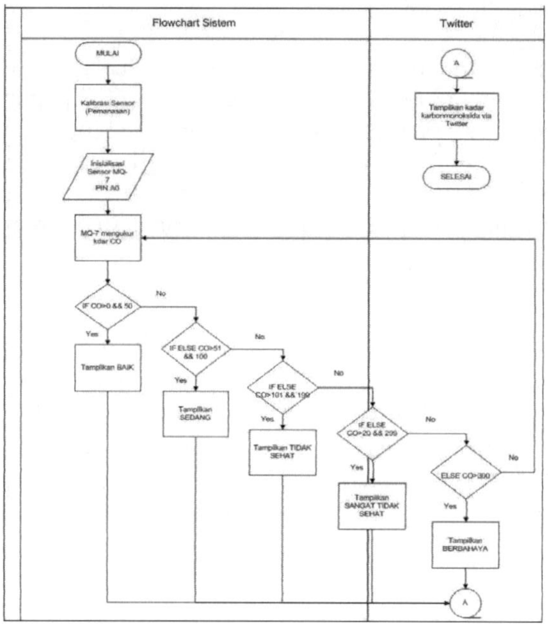

Figure 1. Flowchart System

In this process of integration, component-based software, mechanical design, electronic software, and design are carried out. Assembling is done by uniting all the parts. In this stage programming and configuration between hardware and software is done using the Arduino IDE application. The following is the program or source code compiled using the Arduino IDE application via the Arduino upload board

```
// Ethernet Shield
byte mac[] = { 0xDE, 0xAD, 0xBE, 0xEF, 0xFE, 0xED };
IPAddress dnServer (192, 168, 88, 254);
IPAddress gateway (192, 168, 88, 254);
IPAddress subnet (255, 255, 255, 0);
IPAddress ip (192, 168, 88, 100);
// TOKEN for Log in to Twitter
Twitter twitter ("212564141w5J6efG8aoZZcQ5zD2gyHxjTcRz60vI81BiAzlJX);
char tweetText[140];
void setup()
{
 delay(100);
 Ethernet.begin(mac, ip, dnServer, gateway, subnet);
}
```

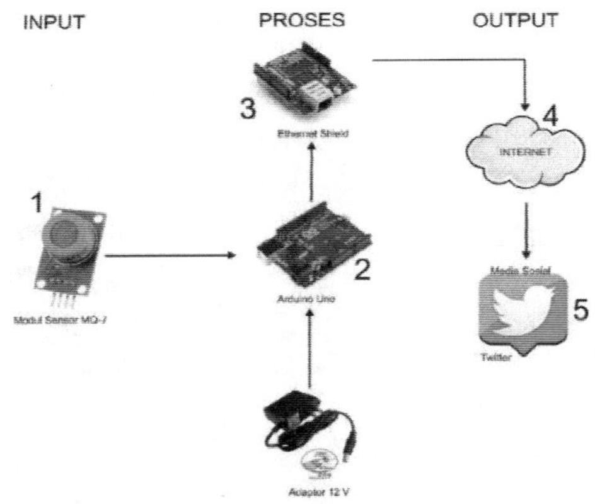

Figure 2. System configuration flow

This sensor will cover gas resistance in the air. The higher the gas reserve in the air. The output of Arduino is that value of CO gas intensity in the air. Data will be transmitted through an Ethernet shield that is connected to the internet and access the twitter.com to be seen. [20]

4. Result

In this study, the system workflow will be explained from the initial stage to the final stage. At the initial stage, it starts by powering the Arduino Uno, the active power indicator is indicated by turning on the indicator light on the Arduino board, then the MQ-7 sensor will activate and detect carbon monoxide gas in the air. If the sensor detects CO gas, it will be sending data to Arduino Uno. After being processed by Arduino Uno it will be sent via the internet by Ethernet Shield and display on twitter.

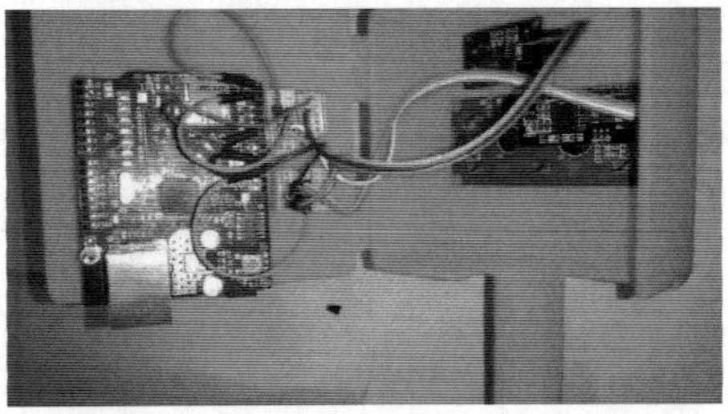

Figure 3. Component Placement

On the Arduino test Uno R3 is done by giving a voltage of 6V - 12V. After that the output voltage is checked on the 5V pin which is connected to positive phobe and the GND pin which is connected to the negative multimeter. This test is done by uploading the program to the Arduino Uno board with the status / tweet display in the form of carbon monoxide levels.

Figure 4. Twitter Status

IOP Conf. Series: Materials Science and Engineering **621** (2019) 012006 doi:10.1088/1757-899X/621/1/012006

4.1. Ethernet Shield Validation

At this stage the MQ-7 Sensor Module is tested as a system input on the A0 pin connected to the A0 pin on the Arduino board, then the value can be seen in the Arduino IDE Serial Monitor. Validation test is carried out by giving motor vehicle exhaust (exhaust) to the system input.

Table 2. Validation of Ethernet Shield

Resistance Value	Respond Time (s)	Result
<300	0.3	Working
>300-400	0.6	Working
>400	0.9	Working

Based on the test data in table 2, can be seen that the sensor response time depends on the presence of vehicle exhaust gas.

4.2. Testing phase

The system testing was carried out for 7 days on the road 5 sub-districts of Bogor City including: South Bogor, East Bogor, West Bogor, Central Bogor, and North Bogor.

a. On March 28, 2016, a study was conducted in East Bogor precisely in Baranang Siang area (in front of KFC).
b. On March 29, 2016 a research was conducted in South Bogor precisely in Heroes' / Empang's jln.
c. On March 30, 2016, a study was conducted in West Bogor precisely at Jl. Batu Batu.
d. On March 31, 2016, a study was conducted in North Bogor precisely at jln. KH. By Iskandar.
e. On April 1, 2016, a study was conducted in Bogor Tengah precisely at jln. KH. Sholeh Iskandar / SAMSAT Office.
f. On April 2 - 3, the study was conducted again in East Bogor and North Bogor, this study aimed to determine the value of monoxide gas on Saturday and Sunday or the increasing number of vehicles on weekends. East Bogor is the main entrance to Bogor City and North Bogor is the exit of Bogor City.

The test results are obtained in the form of graphs which have been averaged as follows:

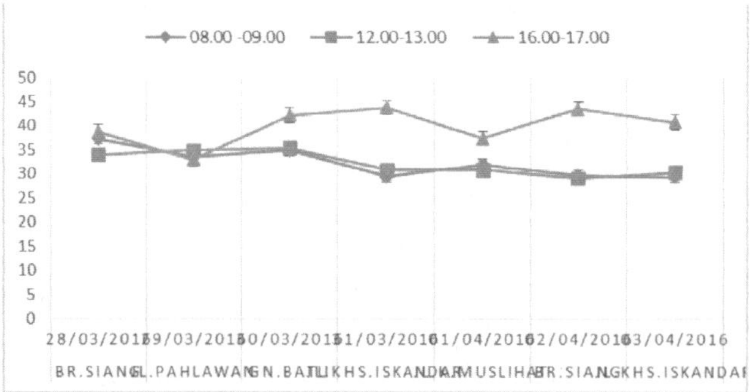

Figure 5. Graph of CO gas levels in Bogor City.

IOP Conf. Series: Materials Science and Engineering **621** (2019) 012006 doi:10.1088/1757-899X/621/1/012006

5. Conclusion

The conclusion of this study is the Arduino Uno-based Real Time Index of Air Pollution Standards (ISPU) through Media Social (Twitter). this system uses Arduino Uno, and Ethernet Shield. The system input uses the MQ-7 Module which will be displayed via the internet using the ethernet shield to the Twitter status. MQ-7 module needs to warm up for 2 hours to initialize the sensor. Tests conducted in this study, the data generated from the study differed (not constant). The number of vehicles and congestion affect the increase in CO gas levels from the same place, the system designed in this study is still manual operation. Based on testing, if the test time is 1 minute, while the tweet is programmed once every 10 seconds, then the tweet that will be displayed is 6 sentences or commonly called 6 tweets.

Tests carried out on several roads in the city of Bogor, the highest CO 2 gas levels were found at Baranangsiang afternoon road with a CO gas value of 43.51939444 ppm with Normal Category on Saturday 2 April 2016 at 16.00-17.00, while the highest data was after the System was enacted. One Direction occurs in Baranang Siang Street, East Bogor with an average value of 42.56527222 ppm, this value is still categorized as Normal.

6. Suggestion

The Real Time Index of Arduino-Based Bogor City Air Pollution Standardization (ISPU) Through Social Media (Twitter) is still not perfect, so it needs improvement. Some suggestions that can be developed in the design network using wireless sensor network topology [19], besides being able to use social media twitter this system can be applied through interfaces such as websites, because in making this system using an Ethernet shield that can store its own data.

7. Acknowledgments

Authors wishing to acknowledge assistance or encouragement from colleagues, special work by technical staff or financial support from organizations should do so in an unnumbered Acknowledgments section immediately following the last numbered section of the paper.

References

[1] Arintya Rahmi, Dian. 2013. Makalah Tugas Akhir. Monitoring Kandungan Karbondioksida (CO2) Dalam Sebuah Model Ruangan Berbasis Mikrokontroler ATmega8535.

[2] Aziz M, Alatas H, Seminar KB. Rancang Bangun Sistem Akuisisi Data Berbasis Wireless Sensor Network (WSN) dan Internet Access Untuk Pengukuran Indesk Standar Pencemar Udara (ISPU) [disertasi]. Bogor (ID): Institut Pertanian Bogor. 2012.

[3] B. Siregar, A. B. A. Nasution and F. Fahmi 2016 "Integrated pollution monitoring system for smart city," *2016 International Conference on ICT For Smart Society (ICISS)*, Surabaya, 2016, pp. 49-52.

[4] Bogor, P. K. 2018. Book 2 Masterplan Smart City of Bogor.

[5] Collins K, Shiffman S, and Rock J. 2016 "How are scientists using social media in the workplace? PLoS ONE"11: e0162680.

[6] DLLAJ 2015 Penyusunan Grand Design Transportasi Kota Bogor. Pemerintah Kota Bogor

[7] G. Anastasi, M. Conti, M. Di Francesco, A. Passarella 2009 Energy Conservation in Wireless Sensor Networks; A survey. *Ad Hoc Networks.* 2009; 7: 537-568.

[8] Haikal Islam, Jilly 2013 Rancang Bangun Alat Pendeteksi Gas CO,CO2, dan SO2.

IOP Conf. Series: Materials Science and Engineering **621** (2019) 012006 doi:10.1088/1757-899X/621/1/012006

[9] Hanwei Electronics. 2002 Technical Data MQ-7 Gas Sensor. https://www.sparkfun.com/datasheets/Sensors/Biometric/MQ-7.pdf (March 9, 2016).

[10] Kadir, A. 2013 Panduan Praktis Mempelajari Aplikasi Mikrokontroler dan Pemrogramannya menggunakan Arduino, Andi Yogyakarta.

[11] Kasar AR, Khemnar DS, Tembhurnikar NP. 2013 WSN Based Air Pollution Monitoring System. *International Journal of Science and Engineering Application.* 2(4): 55-93.

[12] L Linou, Natalia, et al 2018 "Air pollution moves up the global health agenda." (2018): k4933.

[13] Li F, Wang Y, Li X-Y, Nusairat A, Wu Y. 2008 Gateway Placement For *Throughput* Optimization In Wireless *Mesh* Networks. *Mobile Netw Appl,* 13: 198-211.

[14] McRoberts, Michael 2015 Arduino Básico. 2a Edição. Novatec Editora. São Paulo, SP. ISBN: 978-85-7522-404-5.

[15] Mishra S.A, Tijare D.S, Asutkar G.M. 2011 Design Of Energy Aware Air Pollution Monitoring System Using WSN. *International Journal of Advances in Engineering & Technology.* 1(2): 107-116.

[16] Nebath, Everth. 2014. Rancang Bangun Alat pengukur Gas Berbahaya CO dan CO2.

[17] Putut L.E., Widodo Basuki 2011 Simulasi Model Dispersi Polutan Karbon Monoksida di Pintu Masuk Tol. Seminar Nasional Penelitian, Pendidikan dan Penerapan MIPA. Yogyakarta, tanggal 14 Mei 2011. Universitas Negeri Yogyakarta.

[18] United States, Department of Labor . OSHA Occupational Chemical Database. Accessed online 30 January 2016.

[19] V.C. Gungor, Lu B, G. P. Hancke 2010 Opport*uniti*es And Challenges Of Wireless Sensor Networks in Smart Grid. *IEEE Transactions on Industrial Elecronic.* 2010; 57(10).

[20] Yoga Pamungkas, Harly 2015 Alat monitoring Kelembaban Tanah Berbasis Mikrokontroller Atmega168 Dengan Tampilan Output Jejaring Sosial Twitter untuk Pembudidaya dan Penjual Tanaman Hias Anthurium.

IOP Conf. Series: Materials Science and Engineering **621** (2019) 012007 doi:10.1088/1757-899X/621/1/012007

Comparative Analysis of Efficient Image Segmentation Technique for Text Recognition and Human Skin Recognition

Septian Cahyadi[1], **Febri Damatraseta**[2], **Victor Ilyas Sugara**[1]

[1]Faculty of Mathematic and Natural Science Pakuan University of Bogor, Indonesia
[2]Department of Computer Science, Faculty of Master, Budi Luhur Universit, Jakarta, Indonesia.

Corresponding author: septian.cahyadi@unpak.ac.id

Abstract. Computer Vision and Pattern Recognition is one of the most interesting research subject on computer science, especially in case of reading or recognition of objects in real-time from the camera device. Object detection has wide range of segments, in this study we will try to find where the better methodologies for detecting a text and human skin. This study aims to develop a computer vision technology that will be used to help people with disabilities, especially illiterate (tuna aksara) and deaf (penyandang tuli) to recognize and learn the letters of the alphabet (A-Z). Based on our research, it is found that the best method and technique used for text recognition is Convolutional Neural Network with achievement accuracy reaches 93%, the next best achievement obtained OCR method, which reached 98% on the reading plate number. And also OCR method are 88% with stable image reading and good lighting conditions as well as the standard font type of a book. Mean while, best method and technique to detect human skin is by using Skin Color Segmentation: CIELab color space with accuracy of 96.87%. While the algorithm for classification using Convolutional Neural Network (CNN), the accuracy rate of 98%.

1. Introduction

Object detection is a topic of Computer Vision and one of the most interesting research subject on computer science, especially the real-time reading of the camera device. Object detection itself is the first step for the development of advanced research such as the introduction of text and skin for various purposes. The first segmentation is to select the part to be extracted, by detecting the exact object to be processed. In 2015 International Conference of Document Analysis and Recognition (ICDAR) report, on Text on Video challenge, the highest result of text recognition on video has not reached 50% or more precisely 45.18% for Text Localization and 41.84% for End -to-end Recognition. In its development there has been the application of the latest method to answer the challenge of Text Recognition in the video until reaching presentation level of 60%, although there has been a significant increase but seen from the value of the percentage still allows the application of other methods better. Unlike the results of the challenge to read text on still-image both focused text and incidental text, the percentage obtained has reached more than 90%. This is because the scene text is very difficult to read and recognize, because the text that appears on the video is affected by uneven lighting, textured background image or the slope of the position when the video is taken. Because of this, reading texts in the video is a challenge for this researchers.

For object detection to recognize patterns on human skin have different levels of difficulty, human skin has a special color and can be translated into a form of human skin color space. In general, human skin color can be seen on the face, neck, hands and feet. Some studies have successfully analyzed or created a pixel-based skin space in both image and video, including RGB color space, HIS / HSV, TSL, YcbCr, etc. Several studies have conducted experiments on this technique such as, detect the gesture on a face in a video with different light conditions against different backgrounds, which will later be stored into memory [5]. The skin color space can also be used to detect gestures from the hand by extracting features on the hand to merduce noise on complex backgrounds with different light conditions, the way used by selecting vectors from pixels in blocks around the skin tone to obtain skin pixels the new one.

Technology of Computer Vision can also be used for people with disabilities such as helping illiterates (buta huruf) and deaf people(tuli) to know and learn about the letters of the alphabet. In this study we are trying to find better methods and techniques to implement this technology. We are trying to find a good method and technique to make an introduction to a text and detect the presence of human skin on a picture or video in real-time. In this study, we set up a framework for finding good methods and techniques to use.

1) Obtain video or image data from persons with disabilities (inputs).
2) Classify each frame on a video or image to find the detected object.
3) Display the classification results that are most likely or closer than the classification score (output).

The following is an identification of the problems that need to be addressed in the search for good methods and techniques:

1) The use of different light intensity and background on each image or video.
2) The position of the camera to capture an object on a picture or video is very influential.
3) Co-articulation: (when a sign is affected by the preceding or succeeding sign).

2. Method

In this study we will try to compare the segmentation used to recognize an object taken from image or video, that object is a text and human skin. Various methods will be presented in this research ranging from segmentation how to detect a text and detect human contours.

To detect an object to recognize text there are some of the most popular methods, based on the journal literature we refer to. The first most popular method is Neural Network, with variations, such as Artificial Neural Network, Convolutional Neural Network, and Recurrent Neural Network. This method generally performs the initial detection by reading the frame, one of which uses a MESR based detector to take the right frame for use and the easiest to read. Then we use the Neural Network method to classify and classify the text for segmentation. After segmentation done then can be used to do word recognition.

While to detect the presence of human skin contours stored in images or video by converting RGB image into the color space model. In addition to using segmentation of skin color can use certain techniques such as calculating the depth of an image and so forth. The use of skin color segmentation is a simple pixel based feature but its capability is very powerful.

3. Result and Discussion

First of all we look for research what has been done related to image processing for text recognition. This search leads us to a paper entitled "Estimating hidden parameters for text localization and recognition" written by Neumann L and Matas J from the university of Prague. In 2016 [34] also re-proposed this technique with improvements by adding syntethic fonts that have been trained for the classification process, the result is a significant improvement. Tests performed using the ICDAR dataset 2013 achieved an accuracy of 76.3% for text localization and for case-sensitive end-to-end text recognition achieved 45.2%. In the ICDAR dataset 2015 the results achieved for end-to-end text recognition of 41.8% and also in SVT achieved 68.1% result for end-to-end recognition.

Dibayan Chakraborty et al (2013) discusses the technique of taking the character or process of scanning a book more quickly using a digital video camera. This method is assumed to be faster process than scanning a book using a scanner engine. The steps in this method are (1) running a feature learning algorithm on a set of image filters captured before training data to study a set of image features, (2) evaluating the convolutional features of the training image. Reduces the number of features that use spatial pooling, (3) trains a linear classification for either text detection or character recognition. Using the ICDAR dataset 2003 this method achieved an average yield of 82.66% accuracy.

In the experiment (Neumann L, 2013) at the ICDAR conference, a new method was introduced: Oriented Stroke Detection was used to segment the text. The end result of this approach shows a significant result compared to the previous study, tested on the ICDAR dataset 2011, the accuracy value obtained for text localization is 72%, but for end-to-end text recognition only get 45.2% percentage.

In a study conducted by (Baoguang Shi et al., 2016) with a new approach, RARE (Robust Scene Text Recognition with Automatic Refrication) with the same journal title received excellent results. Applied to two datasets, SVT and CUTE80, each received an accuracy of 77% and 59.

Result of research related to best text recognition so far obtained by using method of convolutional neural network (CNN). Introduction is achieved by performing a multi-directional classification across the dictionary to obtain potential words. As a result, of the many datasets tested, from Synthetic Fonts, ICDAR03, SVT, ICDAR13, and IIT5k, an average of 93% was obtained for accurate cropped word recognition. But for end-to-end text spotting achieved an average yield of 75%.

Prajit Ramachandran (2016) discusses the implementation of Faster R-CNN algorithm to detect objects on Video. The results of this implementation are presented at ICCV 2015 ImageNet and MS COCO Visual Recognition Challenges Joint Workshop. This study uses a video clip from the ImageNet 2015 Object Detection from Video Challenge. In this journal it was submitted that the author uses VGGNet as the Convolutional backend of Faster-RCNN. VGGNet provides substantial improvements beyond the actors described by Zeiler and Fergus. From the result of set validation, the writer get the accuracy of mAP 53%.

U. Bhattacharya et al (2016) discusses the technique of detecting Scene Text without limitation using a method that is easy to implement. Based on efficient stroke detection of pixels and comparing with surrounding pixels. Stroke-specific keypoints detect efficiently and text fragments sequentially in extracts using local threshold following keypoint properties. The classification method effectively calculates features to eliminate non-text areas. This method supports a wide variety of scripts (Latin, Hebrew, Chinese, etc). The results of stroke detector testing with FASText on ICDAR dataset 2013 showed an accuracy of 82% with an increase in process speed 4 times higher than MSER.

Annmaria Cherian (2016) discusses the application of the Hough Transform method to improve the orientation of the scene image and use efficient character detection and localization techniques. The method used to filter non-text components is SVM. After the filtering process, character recognition is used to recognize the text accurately. The dataset used in this study is the Text-Perpective Scene dataset which gives the image oaoan ads or billboards located on the edge of the highway with a perplexal image is skewed. And the results of the test show that the method used is quite simple and exceeds the existing method.

Lukas Neumann et al (2016) tried to approach the Extremal Regions used in this journal to read Scene Text localization and recognition or referred to as Photo OCR. The highest probability of characters is then selected in the final stages when the context of each character is known. This method is tested on three public datasets. First use ICDAR 2013 dataset and achieve outstanding results in text localization. Both use SVT datasets and the results significantly outperform other methods. The third dataset uses ICDAR 2015, and achieves exceptionally competitive accuracy.

Raja Vikramdeep Singh et al (2016) discusses the implementation of the OCR methodology to help collect incoming vehicle data through toll roads and traffic monitoring in India. Stages in the process include four phases, the first phase is image acquisition, second extraction, third pixelation and segmentation, and fourth is Recognition. The result obtained an accuracy of 98%.

Gokhan Yildirim et al (2016) describes text recognition techniques on natural images using multiclass hough forest methods. This journal offers a technique for detecting and recognizing text in various conditions by searching for words directly without reducing the image to a region or individual character. They dedicate three contributions. And by using the ICDAR 2003 dataset we get outstanding results in word recognition in terms of performance, which is 85.7%.

In the research (Hartanto and Kartikasari, 2016) conducted HGR research using SIBI as the object of his research. The approach used is non-contact approach by utilizing webcam camera that has been embedded, the method used is trying to catch or detect the movement of the hand by using Skin Color Segmentation model YCbCr color space and Back Propagation Neural Network classification to look for patterns on hand movements that are detected at Webcam in real-time. In their experiments, the first step used to perform the HGR are catching the HGR on the frame. The last stage is the process of classification with Back Propagation Neural Network to find the pattern of input and match it with the SIBI pattern alphabet dataset and display the classification results into a text. From the experiments that have been tested, the accuracy is 62.6% of 26 letters of the SIBI alphabet.

A further reference to the research belongs (Garcia and Viesca, 2016), they experimented on HGR to translate American Sign Language (ASL) in real-time. This research is similar to previous research using Skin Color Segmentation to know the existence of HGR. The HSV color space model and the Convolutional Neural Networks (CNN) classification were used in ASL testing. However they use the ASS pattern dataset ILSVRC2012 from letter a to k (without j) that have been issued by Surrey University and Massey University. From the results of the tests they have done, CNN is used to determine the percentage of accuracy of the object under study. Obtained a percentage of 98% for ASL letters from a to e. While from the letters ASL a through k (without the letter j) obtained by 70%.

Translating sign languages was re-examined by (Shinde and Kagalkar, 2015). Marathi's sign language became the object of their research. The use of Skin Color Segmentation with HSV color space model is used to detect human skin pixel contours in the image. While the classification to understand the input image using Euclidean Distance. Accuracy value in detecting 6 letters Marathi Sign Language can reach 30% -100%. However, the time of classification process used is quite time consuming.

Thai Sign Language (TSL) becomes the object of research (Pariwat and Seresangtakul, 2017) to translate sign language into text. The technique used is using SVM technique to get accuracy value in classifying TSL applying Linear Kernel, Polynomial Kernel, and RBF Kernel methods. The accuracy value in detecting 15 TSL letters was 91.20%.

The Arabic Sign Language (ArSL) translator authentication is investigated by (Saleh et al, 2016) using SOFTKINECT sensors to detect the movement of the human hand, while for the classification of the results taken by sensors using PCANet. The accuracy value for the introduction of 28 ArSL reaches 96% -99.5%.

The research belongs (Shweta et al, 2016) trying to translate ASL sign language into a voice or text to speech. Techniques used in detecting HGR simply change the image into binary form by using Geometric operations, Neighborhood and block operations. For classification use Mel Frequency Cepstrum Coefficients provided by Matrix Laboratory. The results achieved in detecting 9 ASL letters obtained an accuracy of 91%.

The Alphabet Bangla Sign Language is researched by (Azher and Chwdhury, 2016) to translate it into text using Skin Color Segementation model HSV color space techniques to detect the presence of human skin contours and classification for recognition using SVM. From 6 letters Bangla Sign Language got an accuracy score of 97.7%.

(George et al, 2016) conducted a study on how to detect an autonium disease in human skin by using the object of research from the image of psoriasis. To know Skin Map (SM) in image use two approaches that is Otsu's Thresholding and Decision Tree (DT) by using Morphological operation. And to know skin on image using Skin Color Segmentation with model YCbCr, HSV, RGB, XYZ and CIELab color

IOP Conf. Series: Materials Science and Engineering **621** (2019) 012007 doi:10.1088/1757-899X/621/1/012007

space. Histogram-based Bayesian classification is used to conduct trials of 100 psoriasis images. Where this classification is so effective to test the methods presented because Skin Probability Maps (SPMs) generated from the color space presented can produce the best SM for psoriasis images. From the test results obtained for color space YCbCr with Otsu's thresh percentage highest in TNR is 98.81% while for DT the largest percentage value is in TNR of 97.61%. The HSV color space of Otsu's thresh percentage is TNN 98.81% and DT is 97.91%. And the CIELab color space of the largest percentage value is in the TPR of 92.49% with Otsu's Thresh and DT of 94.60% in TPR.

The research (Ting and Lin, 2016) attempted to test the CIELab color space technique on 20 images with minimal contrast. They want to prove that using CIELab and Fuzzy Automatic Contrast Enhancement (FACE) techniques can increase the value of the contras better so as to get clearer and better viewed images. From the results of experiments they have done obtained a contrast increase value of 20 images obtained RMSC value of 12.6 to 243.7.

4. Conclusions

From the elaboration of the use of methods and techniques that have been done by some previous researchers, it can be concluded that in detecting an object, the most preferred is the speed in process to produce better and more accurate information.

4.1. Table comparison of Methodology Text Detection

Based on table 1, many discuss about text recognition, the most widely used and accurate method is Convolutional Neural Network with achievement accuracy reaches 93%. Although researchers generally modify the CNN method, resulting in a variety of results, it can generally be argued that the use of CNN is the best method for the initial phase of text recognition and segmentation. While the next best achievement obtained OCR method, which reached 98% on the reading plate number that has the same font on each vehicle, seen from the value it does look very promising, but the results can not be achieved for reading the text in general with various font types. The best results obtained by the OCR method are 88% with stable image reading and good lighting conditions as well as the standard font type of a book, rather than real-time text readings encountered on the road with very diverse variations.

Table 1. A table with result of comparison methodology of Text Detection.

No	Title	Author	Year Publish	Methodology	Result
1 [a]	Text recognition in natural images using multiclass hough forests	Gokhan Yildirim, Radhakrishna Achanta, Sabine Susstrunk	2013	Hough Forest	85.7%
3 [a]	SceneTextReg: A Real-Time Video OCR System	Haojin Yang, ChengWang, Christian Bartz, Christoph Meinel	2016	Convolutional Neural Network (CNN)	82%
4 [a]	An End-to-End Trainable Neural Network for Image-based Sequence Recognition and Its Application to Scene Text Recognition	Baoguang Shi, Xiang Bai	2016	Convolutional Recurrent Neural Network (CRNN)	83.4%

5 [b]	Robust Scene Text Recognition with Automatic Rectification	Baoguang Shi, Xinggang Wang, Pengyuan Lyu, Cong Yao, Xiang Bai	2016	Robust text recognizer with Automatic Rectification (RARE)	59%
6 [c]	Automobile Number Plate Recognition And Extraction Using Optical Character Recognition	Raja Vikramdeep Singh, Navneet Randhawa	2016	Optical Character Recognition Methodology	98%
7 [c]	Automatic Localization and Recognition of Perspectively Distorted Text in Natural Scene Images	Annmaria Cherian	2016	Hough Transform & SVM	72%
8 [c]	Efficient Scene text localization and recognition with local character refinement	Lukas Neumann, Jiri Matas	2015	MSER with Character Refinement	77.1%
9 [c]	Reading Text in the Wild with Convolutional Neural Networks	Jaderberg et al	2016	CNN	AVG: 93%

[a] Research base on Text Recognition by Scene Text Video Footage (dataset)

[b] Research base on Text Recognition by Video Real-Time (Streaming)

[c] Research base on Text Recognition by Image

4.2. Table comparion of methodology Human Skin Detection

From table 2 it is found that a good skin color segmentation model is used, ie CIELab color space. It is better than other color models that with the value of accuracy can reach 96.87%. And for the classification method obtained Convolutional Neural Network has an accuracy of 98%.

Table 2. A table with result of comparison methodology of Human Skin Detection.

No	Title	Author	Publish	Methodology	Result
1	Implementation of skin color selection prior to Gabor filter and neural network to reduce execution time of face detection	Mejda Chihaoui, Akram Elkefi, Wajdi Bellil dan Chokri Ben Amar	2015	Skin Color Segmentation : YCbCr color space dan Sd-Gabor-NN	Execution time (sec): 12.46500 – 77.77300
2	A Sign Language Recognition Approach For Human-Robot Symbiosis	Afzal Hossian, Shahrin	2014	*Skin Color Segmentation* : YIQ	Accuracy : 80%-92%

		Chowdhury, and Asma-ull-Hosna		*Back Propagation Neural Network*	
3	*Android Based Real-Time Static Indonesian Sign Language Recognition System Prototype*	Rudy Hartanto, Annisa Kartikasari	2016	*Skin Color Segmentation : YCbCr color space;* *Back Propagation Neural Network*	Accuracy : 62.6%
4	*Real-time American Sign Language Recognition with Convolutional Neural Networks*	Brandon Garcia dan Sigberto Alarcon Viesca	2016	*Convolutional Neural Networks*	Accuracy : 70%-98%
5	*Thai Finger-Spelling Sign Language Recognition Using Global and Local Features with SVM*	Thongpan Pariwat dan Pusadee Seresangtakul	2017	*SVM Teknik : Linear Kernel Accuracy, Polynomial Kernel, RBF Kernel, Sigmoid Kernel*	Accuracy : 91.20%
6	Arabic Sign Language Fingerspelling Recognition from Depth and Intensity Images	Saleh Alyl, Basma Osman, Walaa Aly dan Mahmoud Saber	2016	*Sensor SOFTKINECT and PCANet*	Accuracy : 96%-99.5%
7	Real Time Two Way Communication Approach for Hearing Impaired and Dumb Person Based on Image Processing	Shweta S. Shinde, Rajesh M. Autee dan Vitthal K. Bhosale	2016	*Mel Frequency Cepstrum Coefficients and* *Geometric operations, Neighborhood and block operations.*	Accuracy : 91%
8	Hand Sign Language Recognition for Bangla Alphabet using Support Vector Machine	Md Azher Uddin dan Shayhan Ameen Chowdhury	2016	*Skin Color Segmentation : HSV color space dan* *SVM*	Accuracy : 97.7%
9	Pixel-based Skin Segmentation in Psoriasis Images	Y. George, M. Aldeen, and R. Garnavi	2016	Skin Color Segmentation : CIELabs color space and Bayesian Classifier	Accuracy : 90%-97%

5. References

[1] Shinde, Amitkumar dan Kagalkar, Ramesh. *Advanced Marathi Sign Language Recognition using Computer Vision*., International Journal of Computer Applications, Volume 118 – No. 13, pp.1-7. 2015.

[2] Hartanto, Rudy dan Kartikasari, Annisa. *Android Based Real-Time Static Indonesian Sign Language Recognition System Prototype*., International Conference on Information Technology and Electrical Engineering (ICITEE), pp. 1-6. Oct. 2016.

[3] Garcia, Brandon., dan Viesca, Sigberto Alarcon. *Real-time American Sign Language Recognition with Convolutional Neural Networks*, Stanford University., Stanford, CA, 2016.

[4] Pariwat, Thongpan. dan Seresangtakul, Pusadee. *Thai Finger-Spelling Sign Language Recognition Using Global and Local Features with SVM*., 9th International Conference on Knowledge and Smart Technology (KST), pp. 116-120. March 2017.

[5] Saleh Alyl, Basma Osman , Walaa Aly dan Mahmoud Saber. *Arabic Sign Language Fingerspelling Recognition from Depth and Intensity Images*., International Computer Engineering Conference (ICENCO), pp. 99-104. 2016

[6] Shweta S. Shinde, Rajesh M. Autee dan Vitthal K. Bhosale., *Real Time Two Way Communication Approach for Hearing Impaired and Dumb Person Based on Image Processing*., International Conference on Computational Intelligence and Computing Research (ICCIC), pp.1-5. 2016.

[7] Md Azher Uddin dan Shayhan Ameen Chowdhury., *Hand Sign Language Recognition for Bangla Alphabet using Support Vector Machine*., International Conference on Innovations in Science, Engineering and Technology (ICISET). Pp 1-4, 2016.

[8] Y. George, M. Aldeen, and R. Garnavi., *Pixel-based Skin Segmentation in Psoriasis Images*., nternational Conference of the IEEE Engineering in Medicine and Biology Society (EMBC). Pp 1352- 1356, 2016.

[9] Po Ting Lin dan Boting Rex Lin., *Fuzzy Automatic Contrast Enhancement Based on Fuzzy C-Means Clustering in CIELAB Color Space*., International Conference on Mechatronic and Embedded Systems and Applications (MESA). Pp 1-10, 2016.

[10] Y. George, M. Aldeen, and R. Garnavi, *Pixel-based Skin Segmentation in Psoriasis Images*, 2016 IEEE 38th Annual International Conference, Engineering in Medicine and Biology Society (EMBC), 2016.

[11] Lukas Neumann, Jiri Matas, *Real-Time Lexicon-Free Scene Text Localization and Recognition*. IEEE Transactions on Pattern Analysis and Machine Intelligence. 2016

[12] Haojin Yang, ChengWang, Christian Bartz, Christoph Meinel, *SceneTextReg: A Real-Time Video OCR System Haojin*, Hasso Plattner Institute (HPI), University of Potsdam, Germany. 2016

[13] Baoguang Shi, Xiang Bai. *An End-to-End Trainable Neural Network for Image-based Sequence Recognition and Its Application to Scene Text Recognition*. IEEE Transactions On Pattern Analysis And Machine Intelligence. 2016

[14] Baoguang Shi, Xinggang Wang, Pengyuan Lyu, Cong Yao, Xiang Bai. *Robust Scene Text Recognition with Automatic Rectification*. IEEE Conference on Computer Vision and Pattern Recognition Robust. 2016

IOP Conf. Series: Materials Science and Engineering **621** (2019) 012008 doi:10.1088/1757-899X/621/1/012008

Design of power bank mobile using solar panel based microcontroller atmega 328

Agus Ismangil[1*], Haddy Prasetya Susanto[2]

[1]Department of Computer Science, Faculty of Mathematics and Natural Sciences, Pakuan University, Indonesia

[2]College Student of Computer Science, Faculty of Mathematics and Natural Sciences, Pakuan University, Indonesia

*Email : a.ismangil.physics@gmail.com

Abstract. The need for electric energy consumption is increasing. So, in search of new energy sources should meet the requirement that is generating a large amount of energy, economic costs, and no negative impact on the environment. One such energy source is solar energy. By using solar panels, solar energy can be converted into electrical energy. The use of solar panels as a source of electrical energy in the power bank, facilitate the charging of batteries when outside the room or when there is no source of electricity. In this experiment experiments using 5volt solar panels that function as a power source, 5000 mAh battery, and microcontroller ATMega 328 that serves as an output that displays the battery indicator through the LCD. This solar panel device becomes more economical and can be developed again by using components that have more power. Test results using 2 pieces of mobile phone can charge for 1 hour and successfully charge the battery of 26% and 22% respectively. With a fully loaded 5000 mAh battery capacity. Charge the battery by drying the power of the solar panel under the hot sun from 9 am to 3 pm.

1. Introduction

Currently the need for electricity consumption is increasing. So, in search of new energy sources should meet the requirement that is generating a large amount of energy, economic costs, and no negative impact on the environment [1]. One such energy source is solar energy [1]. By using solar cell, solar energy can be converted into electrical energy [2]. One of the communication tools that are very popular in the world is the smartphone [3]. But along with the progress of this battery cell phone storage has not been able to offset various applications in use in smartphones [4]. This makes people often bring power banks where they go [5]. But even powerbank does not have power that never runs out, then we need to fill it and can use it again. In addressing these issues, the author tries to design a battery charging device using solar panels, then made a tool entitled "Design Of Power Bank Mobile Using Microcontroller Based On Atmega 328" With the creation of this tool is expected to help for people who forget to charge the battery and often outside the room. By using a cell phone charger from the solar cell of course the problem can be overcome. The scope of this study is limited to the manufacture or assembly of Power Bank Mobile Using Solar Panel Based Microcontroller ATMega 328. Consists of assembly of battery charging device using solar panel and microcontroller ATMega 328[6]. The benefits of this study are as follows:For the community, Ease and help charging the battery when outside the room or when there is no electricity, More efficient because it uses a source of solar light as a source of battery charging power[7]. For the reader, Can know the Design of Power Bank Mobile Android Using Solar Cell, Know how to assembly the battery charging design components using solar cell, Adds insights in related fields[8].

IOP Conf. Series: Materials Science and Engineering **621** (2019) 012008 doi:10.1088/1757-899X/621/1/012008

2. Literature review

2.1 Solar panels

Solar panels are devices consisting of solar cells that convert light into electricity [1]. They are called solar over the Sun or "sol" because the Sun is the strongest source of light that can be utilized. Solar panels are often called photovoltaic cells, photovoltaic can be interpreted as "light-electric"[2]. The solar cell or PV cell depends on the photovoltaic effect to absorb the Sun's energy and cause the current to flow between two opposite charged layers [3].

2.2 Microcontroller ATMega 328

Arduino Uno is an ATmega328 microcontroller based circuit board. This integrated circuit has 14 digital inputs / outputs (6 outputs for PWM), 6 analog inputs, 16 MHz ceramic crystal resonators, USB connections, adapter sockets, ICSP header pins and reset buttons. This is what is needed to support microcontroller easily connected with USB power cable or AC adapter power supply cable to DC or also battery [4].

2.3 Battery

A device that can convert its stored chemical energy into electrical energy that can be used by an electronic device. Electrical energy generated by this battery is the same as accumulator, which is electric alias DC. The amount of electricity generated depends on how large the battery is. Generally, the battery consists of 2 main types of disposable primary batteries, and secondary batteries that can be refilled [5].

2.3.1 Primary battery (single use battery)

This primary battery or disposable battery is the most common battery on the market, due to its wide usage and its more affordable price. These types of batteries generally provide 1.5 Volts of voltage and consist of various sizes such as AAA (very small), AA (small), C (medium), and D (large). In addition, there is also a box-shaped battery with a voltage of 6/9 Volt [6].

2.3.2 Secondary battery (rechargeable / rechargeable battery)

Secondary battery is a type of battery that can be refilled or rechargeable battery. In principle, the way the secondary battery generates an electric current is the same as the primary battery. However, the chemical reaction of this secondary battery can be reversible. The types of rechargeable batteries we often find include Ni-cd (Nickel Cadmium) batteries, Ni-MH (Nickel Metal Hydride) and Li-Ion (Lithium Ion) batteries [7].

2.4 LCD (Light Crystal Display)

One type of electronic display made with CMOS logic technology that works by not producing light but reflecting light around it against the front-lit or transmitting light from the back-lit. LCD (Liquid Cristal Display) functions as a data viewer either in characters, letters, numbers or graphics [8].

3. Method

In planning a research project required a theme for research can run smoothly and not out of the path that has been set. The theme is "Design of Power Bank Mobile Using Solar Panel Based Microcontroller ATMega 328". Literature is a very useful reverence in the research of a project. Therefore the authors have obtained various literature studies to obtain additional information about the project to be made. Planning a research project includes all the things that will be implemented in research, namely: Determining Research Topics, Estimated tools and materials, Budget estimates, Estimated time, Possible applications and applications to be designed.

IOP Conf. Series: Materials Science and Engineering **621** (2019) 012008 doi:10.1088/1757-899X/621/1/012008

3.1 System Analysis

Phase analysis system is the stage of analyzing all the basic materials that will be used in research. All data that has been collected is analyzed to facilitate the work, such as:

- Microcontroller
- Solar Panel
- Thermoelectric Cooler
- Solar Panel Controller
- And other supporting materials

3.2 Coding / Programming

Coding is a programming component in which the programmed component is Arduino Uno. Here's a flowchart like the following picture:

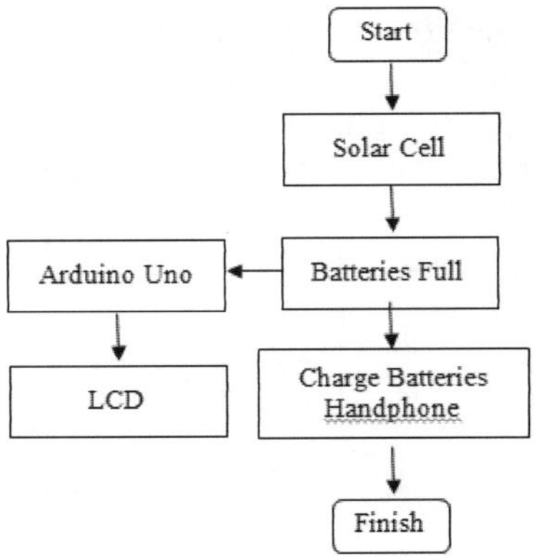

Picture 1. Flowchart process

4. Results And Discussion

At the design stage and the previous implementation has been explained about the power bank design of microcontroller based on ATMega 328. The source code is input into the Arduino as a command for the tool to work properly. Here is the source display in the Arduino IDE app. Here's the source code:

```
Const float referenceVolts = 5.0; //
Const int batteryPin = 0; //
Void setup ()
{
Serial.begin (9600);
}
Void loop ()
{
Int val = analogRead (batteryPin);
Float volts = (val / 1023.0) * referenceVolts;
Serial.println (volts);
Delay (500);
}
```

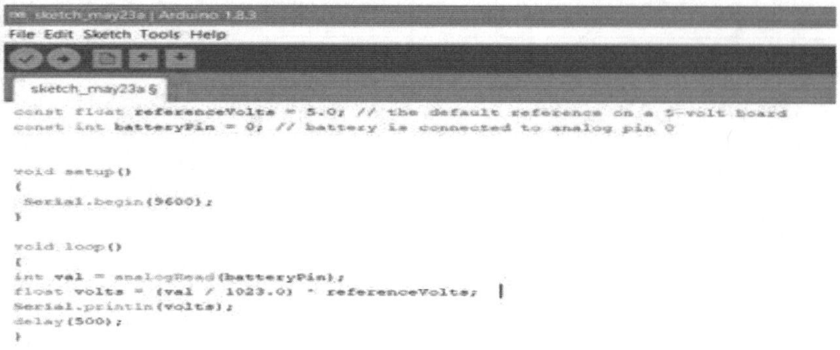

Picture 2. Arduino IDE Display

Serial.begin (9600) This program is used for communication between Arduino with Computer or other devices using Serial communication line, Float volte = (val / 1023.0) creates a variable named voltage in the form of float, in order to generate a voltage value, Serial.println (voltage) displays the value in Serial Monitor, which is stored on the voltage variable. A good program should have an efficient way of working. To facilitate the making of the program then made a flowchart to get the steps sorted. Here is a flowchart for Arduino programming. Here's the picture:

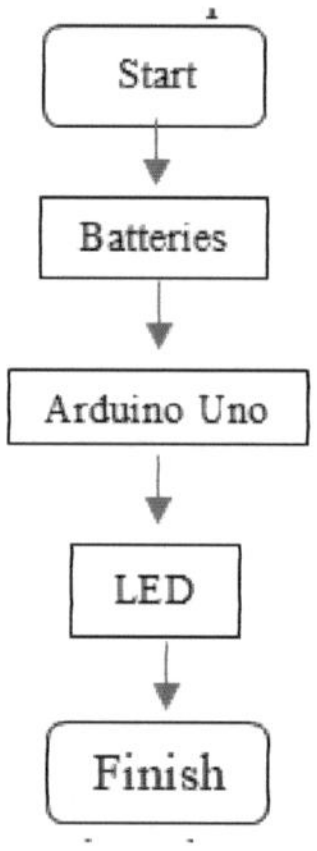

Picture 3. Flowchart Program

The discussion in this chapter is about testing each component of the pliers used and testing the whole system, whether it is running properly or not. The test stages are done with structural testing, functional trials and validation trials. Structural testing is done to ascertain whether all components are properly connected or not and also ensure that all components are structured. Here are the results of structural testing: The arduino and batteries are connected by cables on pins that are already available in each component. Here's the look in the picture:

4.1 Functional Trial

Arduino Uno module checking is done by connecting a 3.7 volt battery in the Arduino module. There is an indicator indicating that the Arduino module is on.

IOP Conf. Series: Materials Science and Engineering **621** (2019) 012008 doi:10.1088/1757-899X/621/1/012008

Picture 4. The green light on the arduino is on

4.2 LCD Testing

LCD checking is done by connecting through arduino. under normal circumstances there is an LCD can be lit. The LCD is marked with a red led indicating it is working.

Picture 5. The lights on the LCD indicate that it is working

4.3 Validation Test

The validation test is a test of the program and the performance of the tool that has been created. In this test is checked for all the functions and features of the tool. mualai from the sensor check until the input and output checks are used. Here are the results of the trial. This test is done by entering usb cable through usb port to handphone, indicator of usb lamp lights indicating power bank can be used. Here's the picture:

Picture 6. USB Indicator

4.4 Arduino Testing

This test is done by giving input to the LCD to display the battery voltage indicator. This test is conducted to determine whether the tool is made to work in accordance with the program that has been made. Here's the picture :

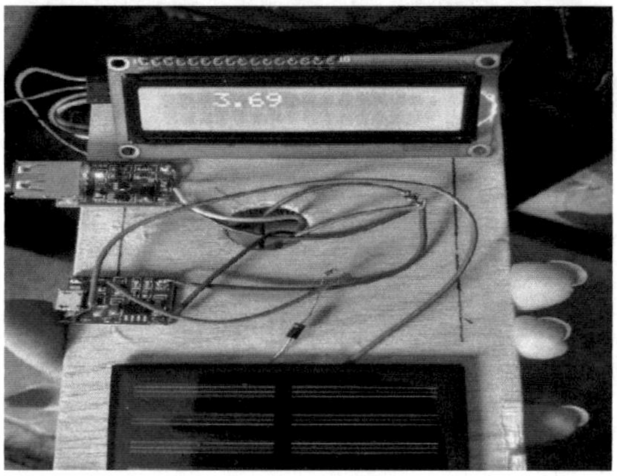

Picture 7. The LCD displays the battery voltage

4.5 Test Results

Test results using 2 pieces of mobile phone with a time charger for 1 hour and managed to charge each battery is 26% and 22%. With a capacity of 5000 mAh battery that is fully charged. Charging the battery by drying the solar panel bank power under the hot sun from 9 to 3 pm. Here is the description in the table:

Table 1. Test Result

Handphone	Capacity	Time Charging	Ful Charging
Xiaomi Redmi 3	4100 mAh	1 jam	26%
Xiaomi Redmi 4A	3120 mAh	1jam	22%

5. Conclusion

Based on the results of this study can be concluded that this tool can work optimally if supported by natural factors that support such as sunny weather and heat from sunlight enough, for charging the source of electricity from solar panels can work optimally. Solar cel used is not too big so the charging process becomes old, can be developed again by using solar panels that have more power than current. The series used works well, able to charge the phone. The LCD circuit that is used as an indicator of the illuminated LCD battery has also displayed a voltage indicator on the battery. And the state of the battery when the tool stand by can also last long. Test results using 2 pieces of mobile phone can charge for 1 hour and successfully charge the battery of 26% and 22% respectively. With a fully loaded 5000 mAh battery capacity. Charge the battery by drying the power of the solar panel under the hot sun from 9 am to 3 pm.

References

[1] Ima maysha 2013 Pemanfaatan Tenaga Surya Menggunakan Rancangan Panel Surya Berbasis Transistor 2N3055 dan Thermoelektrik Cooler

[2] Hendra Maryanto 2010 Analisis Perbandingan Penggunaan Solar Cell Terpusat Dengan Solar Cell Terdistribusi Untuk Memenuhi Kebutuhan Energi Listrik Pada Ruang Kuliah Lantai 4 Gedung FTI

IOP Conf. Series: Materials Science and Engineering **621** (2019) 012008 doi:10.1088/1757-899X/621/1/012008

UII

[3] Peng H 2014 Miniature wire-shaped solar cells, electrochemical capacitors and lithium-ion batteries.

[4] Ryan Ardi Winata 2018 Prototype Kendali Otomatis Penerangan Taman Dengan Tenaga Surya Berbaasis Arduino.

[5] Khalid Fadhullah 2017 Solar Tracking System Berbasis Arduino.

[6] Yossie Widiatmoko 2013 Prototype Pemanfaatan Solar Cell Sebagai Sumber Energi Pada Sistem Otomatisasi Lampu Penerangan Taman.

[7] Susiana Silangit 2009 Penggerak Panel Surya Menggunakan Sensor Cahaya (LDR) Dengan Pengendali Mikrokontroler ATMega 8535.

[8] Heryanto, DKK 2008 Mikrokontroller ATMega 8535

IOP Conf. Series: Materials Science and Engineering **621** (2019) 012009 doi:10.1088/1757-899X/621/1/012009

Stepwise Approach in Lagged Variables Time Series Modeling: A Simple Illustration

Yusma Yanti[1], **Septian Rahardiantoro**[2]

[1]Department of Computer Science, Pakuan University, Indonesia
[2]Department of Statistics, Bogor Agricultural University, Indonesia

Corresponding author: yusmayanti.fn@gmail.com

Abstract. Modeling approach in time series data commonly used to forecast the response variable based on the pattern of the predictor variables. The complicated cases occurred when in the model need the lagged variables from these variables. It can increase the number of predictor variables in the model. In this research, the increasing of number of predictor handled by the stepwise method in the regression analysis approach. All possible lag from the variables generated before the stepwise take action to choose the appropriate variables in the model. The illustration in this research based on the advertising-sales relationship of 36 consecutive months. Result of the model can determine the significance predictor variables in the model, and also can give the appropriate goodness of fit criteria for forecasting.

Keywords: forecasting, lagged variables, stepwise regression, time series regression.

1. Introduction

Time series modeling is a field of research that has the purpose of collecting and studying past observations over a period of time to develop appropriate models that describe the inherent structure of the series. The model obtained will be used to make a series forecast in the future [1]. Forecasting the time series can thus be called the act of predicting the future by understanding the past. Forecasting is very important to predict the future in some areas such as business, economics, finance, science and engineering, etc[2]. Because of this, in order to obtain the correct forecasting and in accordance with the expected model, it should have sufficient data for the time series underlying it. It is clear that successful time series forecasting depends on constructing the right model. In a lot of cases of time series modeling, the lagged variables of predictors or the response often needed to construct the model. In this situation can make the number of predictor variables in the model increase. Therefore, the model constructed become more complicated and difficult to understand. The method for reducing these predictor variables nacessary to take in this case one of the appropriate method is stepwise approach.

Stepwise regression is one of the methods to get the best model from a regression analysis. The first step is done by determining the correlation matrix between the dependent variable with the independent variable. The first variable entry is the variable with the highest correlation and significant with the dependent variable, the second incoming variable is the variable whose partial correlation is highest and still significant, after a certain variable goes into the model then the other variables in the model are evaluated, if any variable which is not significant then the variable is issued [1].

This research will apply the stepwise process to reduce the number of predictor variables from all the possible lagged predictor variables in the model. Data in this research reported in Blattberg and Jeuland (1981), that is advertising-sales relationship data cases. In the end of this research also obtain the

IOP Conf. Series: Materials Science and Engineering **621** (2019) 012009 doi:10.1088/1757-899X/621/1/012009

forecasting from the final model, and the evaluating how good the model constructed to forcast the sales variable.

2. Methods

The approach of this research would be determined in the process below: assume there are a response variable (y_t)and an independent variable (x_t), therefore the regression model that can construct [3]:

$$y_t = \beta_0 + \beta_1 x_t + \varepsilon \quad (1)$$

1. Generate all possible of lagged variable from response and predictor variable to be the predictor variables in the model. Let k number of lag to generate, therefore the predictor variables generated such as:

$$y_{t-1}, \dots, y_{t-k}, x_{t-1}, \dots, x_{t-k}$$

2. Construct the regression model using all lagged variables and predictor variable:

$$y_t = \beta_0 + \beta_1 x_t + \beta_2 x_{t-1} + \dots + \beta_{k+1} x_{t-k} + \beta_{k+2} y_{t-1} + \dots + \beta_{2k+1} y_{t-k} + \varepsilon \quad (2)$$

3. Apply the stepwise process using specified alpha to enter and alpha to remove.
4. Construct the regression model based on the predictor variables selected in the step 3 and evaluate the goodness of fit from these model.
5. Take the forecasting based on the model constructed in the step 4 and evaluate the forecasting goodness of fit using $MAPE$ and $RMSE$ with formula [4]:

$$MAPE = \frac{1}{n}\sum_{t=1}^{n}\left|\frac{e_t}{y_t}\right| \times 100\% \quad (3)$$

$$RMSE = \sqrt{\frac{1}{n}\sum_{t=1}^{n} e_t^2} \quad (4)$$

3. Data

The data in this research consist two variables with 36 consecutive months, which the study focus on the relationship between advertising and sales, reported in Blattberg and Jeuland (1981). The variables are 36 consecutive monthly sales (y_t) and advertising expenditures (x_t) of a dietery weight control product. Researchers in marketing have realized that the effect of advertising may not give effect in the same period in which the advertisement is seen.Therefore, the lagged variables needed in this model from the base model form:

$$y_t = \beta_0 + \beta_1 x_t + \varepsilon$$

For the process of constructing the model, the 36 series of data seperated to be two parts. First part consists 30 series that called training data for constructing the best model. After the best model constructed, this model evaluated in the remaining 6 series of data that called testing data. The evaluation of forecasting by $MAPE$ and $RMSE$ using the testing data.

4. Result and Discussion

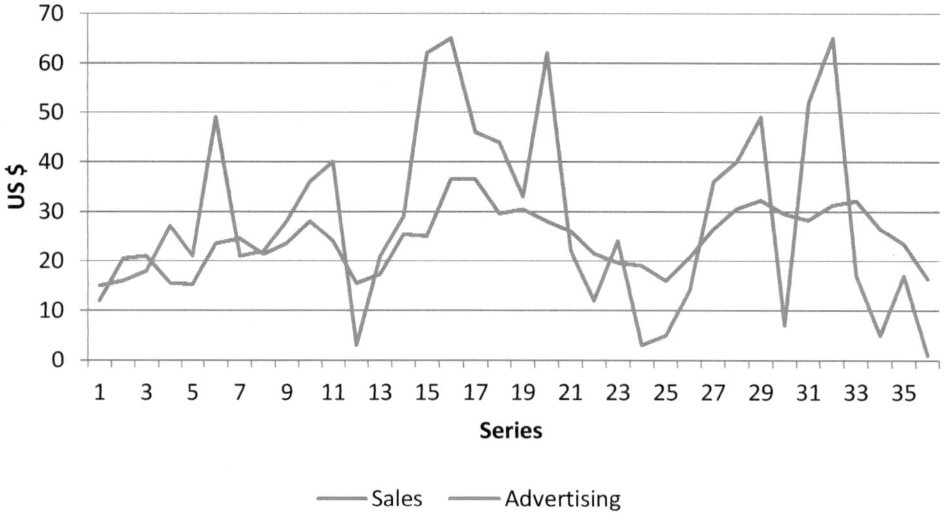

Figure 1. Time series plot of the data

Based on Figure 1 in general, it can be stated that with the increase in advertising costs, then sales will increase as well. The value of sales experienced irregular ups and downs from the first month to the 36th month. A very significant increase in sales occurred in the 12th month to the 17th month. But after that there was a long decline and then there was a spike in purchases again after the 25th month, although the value of sales did not reach the maximum value that occurred in the month-to- 16. When viewed from the graph of advertising costs, overall advertising costs used in a period almost doubled the cost of sales. High enough ads look greatly affect sales, this is evidenced by the maximum ad month-16 and the maximum sales at the same time.

For modeling purpose, lag of sales and advertising value need to be generated, in this case lag 1 until lag 6. Figure 2 and 3 describe the correlation value between sales and lag of sales, also lag of advertising. Correlation calculation results in Figure 2 obtained that the largest correlation between sales with sales lag 1 that is equal to 0.686. The value can interpret that between sales with sales lag 1 there is a considerable positive correlation. In other words, sales value of this time has strong relationship with the sales value in the one period past. As viewed overall, the correlation continues to decrease as the calculation increases with the next lag sale. In lag 6 sales there is a negative correlation value of -0.365.

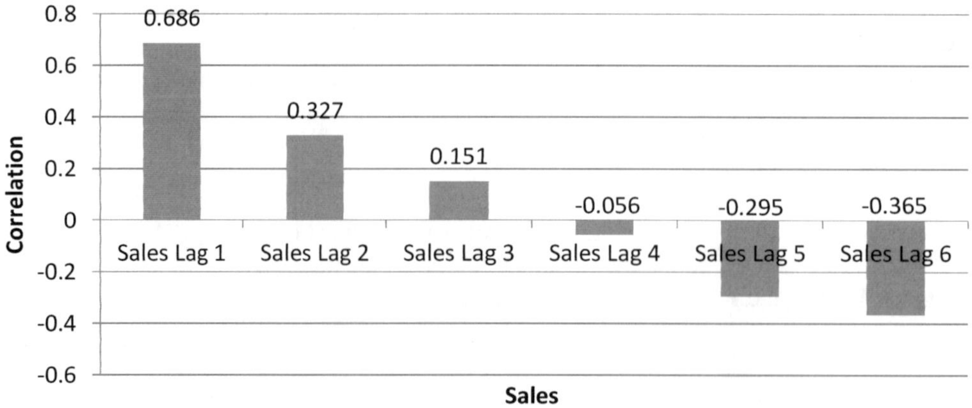

Figure 2. Correlation sales with lag of sales

Figure 3 demonstrates the correlation value between the sales and lag of advertising. It can be seen that the highest correlation occured between sales with advertising lag 1, 0674. The correlation value decreased along with the addition of advertising lag based on this figure. The big enough of positive correlation value also obtained between sales with advertisement that is 0.631.

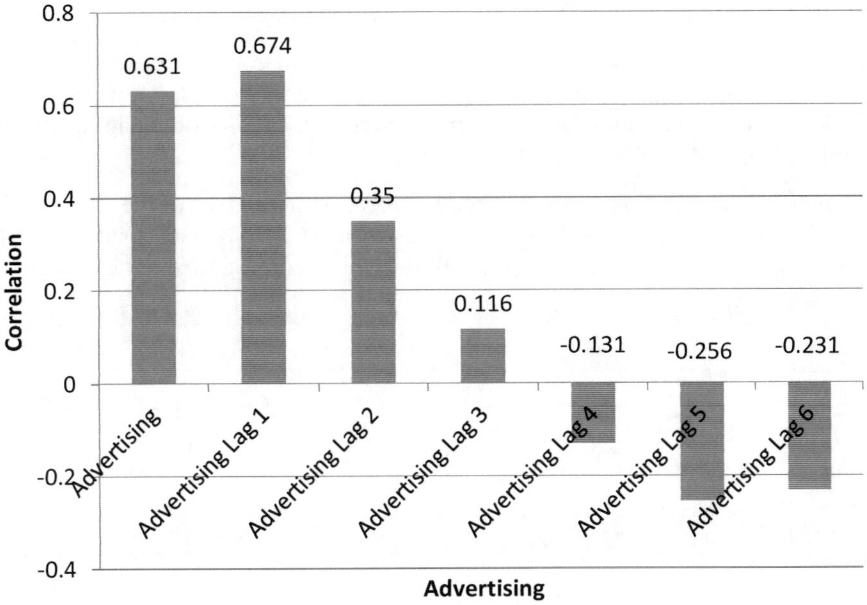

Figure 3. Correlation sales with lag of advertising

The next step, stepwise method applied in the data with the response variable of sales, and all possible predictor variables of lag of sales and advertising. The result of stepwise analysis can be seen in table 1:

Table 1. Stepwise Result

Step	Constant	Variables to enter				Goodness of fit			
		Advertising Lag 1	Advertising	Sales Lag 1	Advertising Lag 5	S	R-Sq	R-Sq(adj)	Mallows Cp
1	19.24	0.203 (0.000)				4.3	46.7	44.8	25.3
2	16.54	0.153 (0.000)	0.145 (0.000)			3.3	68.7	66.41	6.1
3	10.07	0.099 (0.014)	0.132 (0.000)	0.330 (0.020)		3.1	74.7	71.79	2.4
4	12.69	0.105 (0.007)	0.127 (0.000)	0.3 (0.024)	-0.064 (0.044)	2.9	78.6	75.12	0.7

Note: ( ) indicates p-value

In this analysis there are four steps. In the first step, variable enter in the model is advertising lag 1. Step two obtained advertising variables, step three obtained variable sales lag 1, and on step four obtained advertising lag 5. Based on the goodness of the model with the lowest S value of 2.87 and the value R-

Sq (adj) is the highest of 75.12, it can be concluded that this analysis is good enough. So the model chosen is model with predictor of advertising lag 1, advertising, sales lag 1, and advertising lag 5

The following is the result of regression modeling time series with stepwise method with predictor variable based on stepwise method.

Sales = 12.2 + 0.104 Advertising Lag 1 + 0.125 Advertising + 0.320 Sales Lag 1- 0.0621 Advertising Lag 5

Based on these results the four selected variables affect sales significantly. Consequently in this case the sales variable at time t is affected by advertising at time t-1, advertising, sales at time t-1 and advertising at time t-5 with R-Sq of 78.5. Using the model, the goodness of prediction using data testing obtained good results with RMSE of 4.555 and MAPE 13.149%.

In addition, the picture below in Figure 4 presented the forecasting pattern of the analysis. Based on this illustration, the result of regression modeling time series with predictor variable selection using stepwise method can be used as a good alternative to choose variable containing time series data.

Figure 4. Time series plot of forecasting based on the final model

5. Conclusion

Based on the illustration presented in this study, the selection of predictor variables in the regression model using stepwise method on time series data gives a good result. Thus, although many predictor variables are generated based on the time lag, the researcher does not need to be binging to select which predictor variables are in the model. This approach seems good to reduce the number of predictor variables based on lagged variables generated.

References

[1] Wang, Mengchao, Wright J, Buswell R & Brownlee A 2013 A Comparison of Approaches to Stepwise Regression for Global Sensitvity Analysis Used with Evolutionary Optimalization. 13th Conference of International Building Performance Simulation Association, Chambéry, France, August 26-28

[2] Abraham, Bovas & Ledolter J 1981 *Statustical Methods for Forecasting*. John Wiley & Sons, Inc.

[3] Adhikari R & Agrawal K 2013 *An Introductory Study on Time Series Modeling and Forecasting*.

[4] Myers R & Milton J 1991 *A First Course in The Theory of Linear Statistical Models*. PWS-KENT Publishing Company

IOP Conf. Series: Materials Science and Engineering **621** (2019) 012010 doi:10.1088/1757-899X/621/1/012010

The implementation of the Kruskal algorithm for the search for the shortest path to the location of a building store in the city of Bogor

Erniyati[1], P Citra[1]

[1]Department of Computer Science, Faculty of MIPA, University of Pakuan Bogor

Corresponding Author : neni_erniyati@yahoo.com

Abstract. The shortest path is a very important thing in a business sector, especially in sending goods by land. One of the obstacles faced in the shipment of goods is the determination of the trajectory. Determining the best trajectory can minimize shipping costs and time is more efficient A designated algorithm or system is required to search for the shortest and the most optimal traffic path from various numbers of existing alternative pathways. The Kruskal algorithm is one of the algorithms used in solving minimum spanning tree problems by choosing the side that has the smallest weight of each node with the notes that the selected sides cannot form a closed circular. For Case on study, we used data activities is controlled by land transportation from PT. T A. The results showed that with the application of the Kruskal algorithm in the case of finding the closest route to the building store for shipping goods, the optimal path was obtained. The distances of the 12 data that was tried there were 83% data which were smaller than the google map direction API.

1. Introduction

Determination of the trajectory for the delivery of goods is very necessary. in a business sector, especially in sending goods by land. PT T A is a company engaged in the distribution of goods. One of the obstacles faced in the shipment of goods is the determination of the trajectory. Determining the best trajectory can minimize shipping costs and time is more efficient. One of the best pathways is to find the shortest path on the distribution path. The shortest path can be calculated using graph theory. The Kruskal algorithm is one of the algorithms in graph theory in finding the minimum spanning tree to connect each tree in a forest [1]. The basic rule of the Kruskal algorithm is that you have to choose the side that has the smallest weight of each node with the note that the selected sides cannot form a closed circular [2].

The city of Bogor is one of the cities that has very many trajectories or lanes so that to carry out goods delivery requires a lot of time and cost. This paper aims to implement the Kruskal algorithm for the search for the shortest path to the location of a building shop in the city of Bogor and compare the distance obtained by using the Kruskal method and Google map direction Formatting the title, authors and affiliations Please follow these instructions as carefully as possible so all articles within a conference have the same style to the title page. This paragraph follows a section title so it should not be indented.

2. Research Methods

2.1 Graph

The concept of graph theory was first introduced in 1736 to solve the problem of the Königsberg Bridge. The Königsberg bridge problem asks if the seven bridges of the city of Königsberg (left Figure; Kraitchik 1942), formerly in Germany but now known as Kaliningrad and part of Russia, over the river Preger can all be traversed in a single trip without doubling back, with the additional requirement that the trip ends in the same place it began. This is equivalent to asking if the multigraph on four nodes and

IOP Conf. Series: Materials Science and Engineering **621** (2019) 012010 doi:10.1088/1757-899X/621/1/012010

seven edges (right Figure) has an Eulerian cycle. This problem was answered in the negative by Euler (1736), and represented the beginning of graph theory. Graph G is defined as a set of pairs (V, E) written with a notation G = (V, E), which in this case V is a non-empty set of vertices (vertices) and E is a set of connecting edges a pair of knots[3].

a. Trees

A tree is a data structure made up of nodes or vertices and edges without having any cycle. The tree with no nodes is called the null or empty tree. A tree that is not empty consists of a root node and potentially many levels of additional nodes that form a hierarchy. As an example in Figure 1 graph G there are 2 articles, including v1v2v3v6 and v3v4v5, therefore graph G is not a tree. In Figure 1 an example of a number of trees and not trees is given.

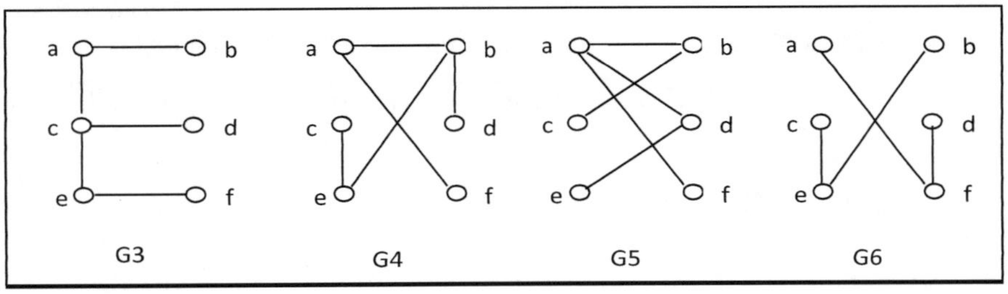

Figure 1. G3 and G4 are a tree, G5 and G6 are not trees

In the example of Figure 1, G3 and G4 are said to be a tree because all the nodes contained in the graph are connected and in G3 and G4 do not contain the particle subgraph. While G5 and G6 are not trees, because even though all vertices in the G5 graph are connected but there are The particle is adf and node in G6 graph is not connected all.

b. Minimum Spanning Tree

A forest is a circuitless graph. A tree is a connected forest. A subforest is a subgraph of a forest. A connected subgraph of a tree is a subtree. A spanning tree of a connected graph is a subtree that includes all the vertices of that graph. Figure 4 is an illustration of a spanning tree T1, T2, T3, and T4 derived from the G7 graph.

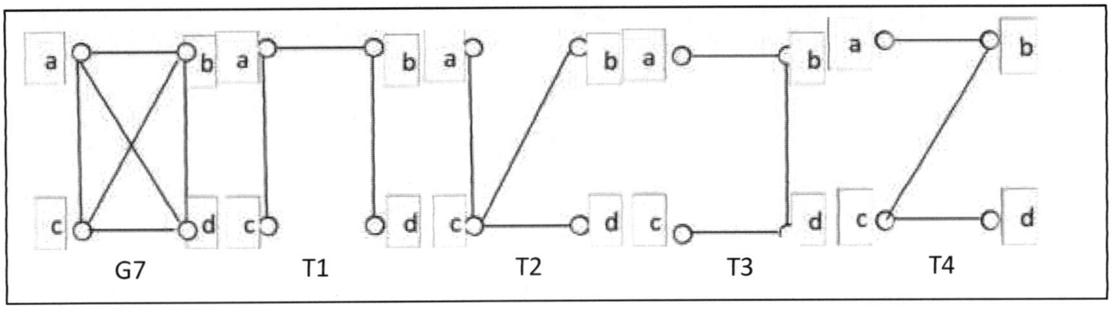

Figure 2. G7 is a graph and T1, T2, T3, and T4 are trees

In Figure 2, it can be seen that G7 is a connected graph and T1, T2, T3, and T4 are spanning tree from G7 graph. The spanning tree has a minimum weight called the minimum spanning tree, as given by Figure 3.

IOP Conf. Series: Materials Science and Engineering **621** (2019) 012010 doi:10.1088/1757-899X/621/1/012010

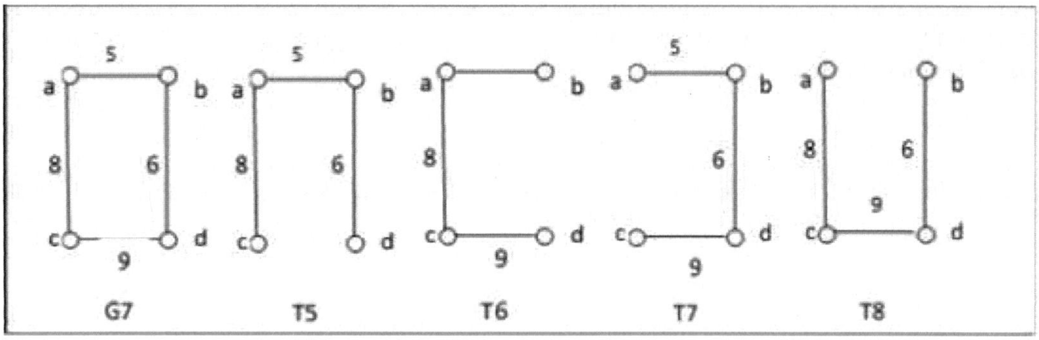

Figure 3. G7 is a simple graph that is connected and weighted,

T5, T6, T7, and T8 is a spanning tree

In Figure 3, it can be seen that G7 is a simple graph that is connected and weighted, that is the weight of each side is 5, 6, 8, 9 and the total weight is 28. The spanning tree T5, T6, T7, and T8 is a spanning tree from G7. The total weight of T5 is 19, the total weight of T6 is 22, the total weight of T7 is 20, and the total weight of T8 is 23. So it can be said that T5 is the minimum spanning tree because the weight of T5 is the smallest weight among the other trees [4].

2.2 Kruskal Algorithm

The Kruskal algorithm is one of the algorithms used in solving the problem of minimum spanning trees that can be applied in various fields of daily life with the aim of optimization. To solve the problem of finding the shortest path using the kruskal algorithm [5] follows the basic rules of the algorithm shown in pseudocode as follows:

```
MST-KRUSKAL(G, w)
A ← Ø
for each vertex v ∈ V[G]
    do MAKE-SET(v)
sort the edges of E into nondecreasing order by weight w
for each edge (u, v) ∈ E, taken in nondecreasing order by weight
    do if FIND SET(u) ≠ FIND SET(v)
        then A ← A ∪ {(u, v)}
             UNION(u, v)
return A
```

2.3 Determination of the cost between two related points

For each node taken from the point of crossing the road to be passed. To determine the cost between two related points is searched by using google maps direction by looking at the distance from point s to n. The result of measuring the value of cost between two interconnected nodes is the cost node to a is 500m, as illustrated in Figure 4.

2.4 Identification of Problems

To determine the closest path to the location of a building store in the city of Bogor Indonesia, by analyzing the data that has been obtained by finding the shortest path from s (the starting point is on Kranggan No. 100 Keranggan Road, Gunung Putri, Bogor) to 12 store locations with through several nodes from the path graph as in Figure 7. The 12 store locations in Bogor are:

p1 (toko Subur Jaya, Jl. Raya Tajur No. 315 Kec. Bogor Timur, Kota Bogor)	p7 (TB. Pandu Makmur , Jl. Tegal Gundil Kec. Bogor Utara)
p2 (TB. Hutan Mas, Jl. Raya Tajur Ciawi Kec. Bo gor Timur)	p8 (TB. Jawa Jl. Raya Bogor Kedung Halang Km. 55 No. 225 Kec Bogor Utara)
p3 (TB. Kurnia Jl. Raya Tajur No. 63A Pakuan, Kec. Bogor Selatan)	p9 (TB. Abadi Jaya Toko, Jl. Sindang Barang No. 61 Gunung Batu, Kec. Bogor Barat)
p4 (TB. Setia Makmur, Jl. Pahlawan No. 183 Bondongan, Kec, Bogor Selatan),	p10 (TB. Sarana Bangunan, Jl. Raya Semplak No. 196 Atang Senjaya Kemang, Kec. Bogor Barat)
p5 (TB. Prapatan Tegal Lega, Jl. Tegal Lega No. 1, Kec. Bogor Tengah)	p11 (TB. Wahana Agung , Jl. Raya Cibadak - Ciampea Cibadak Tanah Sareal Kota Bogor)
p6 (TB. Setia Abadi, Jl. Pengadilan No. 1 DD Kec. Bogor Tengah)	p12(TB. Purba , Jl. Kayu Manis No. 85 Cibadak Kec. Tanah Sareal Kota Bogor)

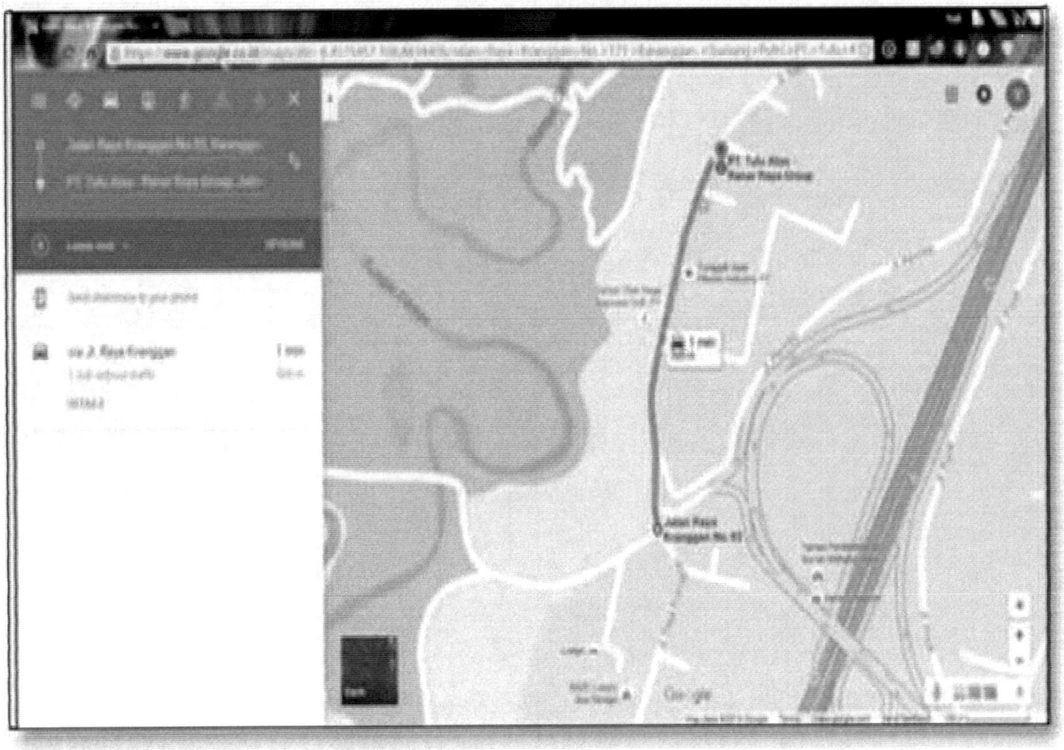

Figure 4. Cost between two Nodes

IOP Conf. Series: Materials Science and Engineering **621** (2019) 012010 doi:10.1088/1757-899X/621/1/012010

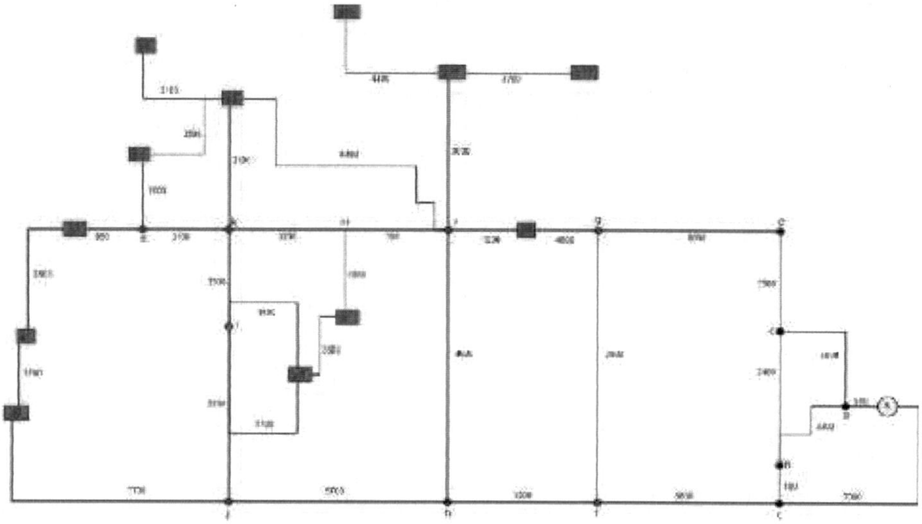

Figure 5. Path Images between Nodes

3. Discussion

Table 1 is a graph data with the distance between vertices, using the Kruskal algorithm step 1 the data is sorted, the results can be seen in table 2. Steps 3 through step 7 can be seen in Figure 8. The results of the path obtained by using the Kruskal algorithm from s to p8 (one of the building stores) are s-a-b-c-f-h-i-p8, the illustration can be seen in Figure 9. The results of testing using the Kruskal algorithm and the results of the calculation of google map direction, from s the 12 building stores can be seen in Table 3.

Table 1. Graph data with each distance between vertices

Nomor	From Verte :	To Vertex	Distances	Nomor	From Verte :	To Vertex	Distances
1	s	a	500	8	d	e	5500
2	s	c	7000	9	e	g	8000
3	a	b	3300	10	f	g	2500
4	a	d	4800	11	f	h	3200
5	b	c	450	12	g	P8	4800
6	b	d	2400	13	h	i	4600
7	c	f	5800	14	i	P8	1200

Table 2. Data that has been sorted by distance

No.	From Vertex	To Vertex	Distances	No.	From Vertex	To Vertex	Distances

IOP Conf. Series: Materials Science and Engineering **621** (2019) 012010 doi:10.1088/1757-899X/621/1/012010

1	b	c	450	8	h	i	4600
2	s	a	500	9	a	d	4800
3	i	P8	1200	10	g	p8	4800
4	b	d	2400	11	d	e	5500
5	f	g	2500	12	c	f	5800
6	f	h	3200	13	s	c	7000
7	a	b	3300	14	e	g	8000

Step 1 (b-c)

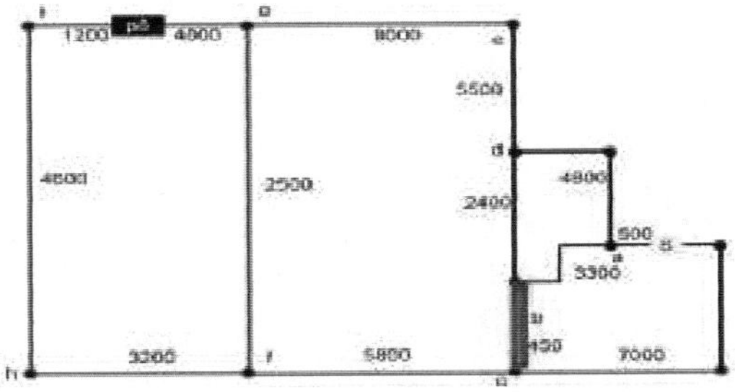

Step 2 (s-a)

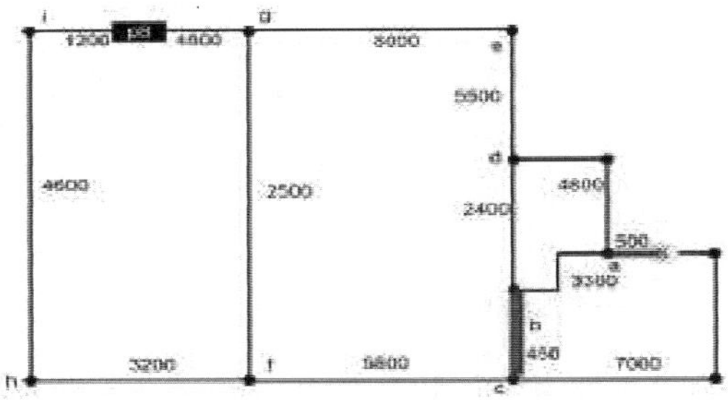

Step 3 (i-p8)

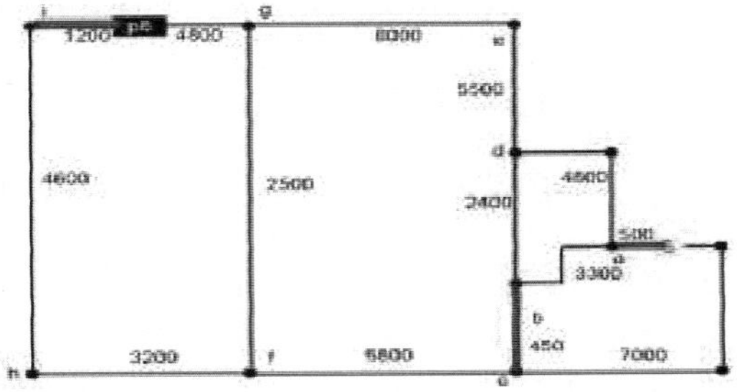

Step 4 (b-d)

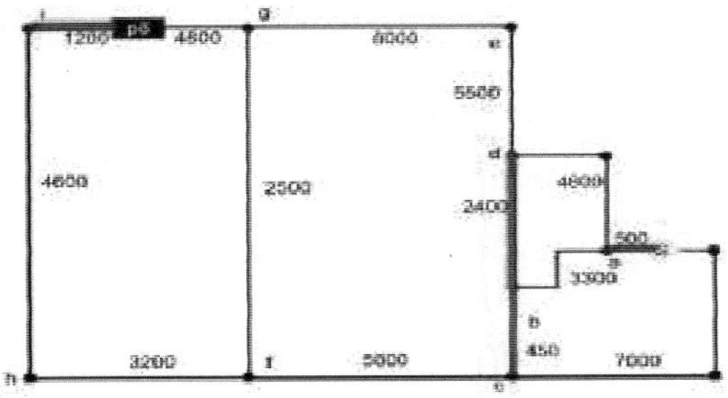

Step 5 (f-g)

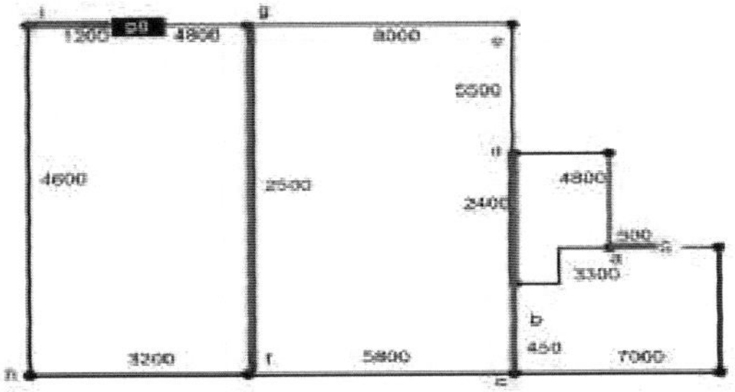

Step 6 (f-h)

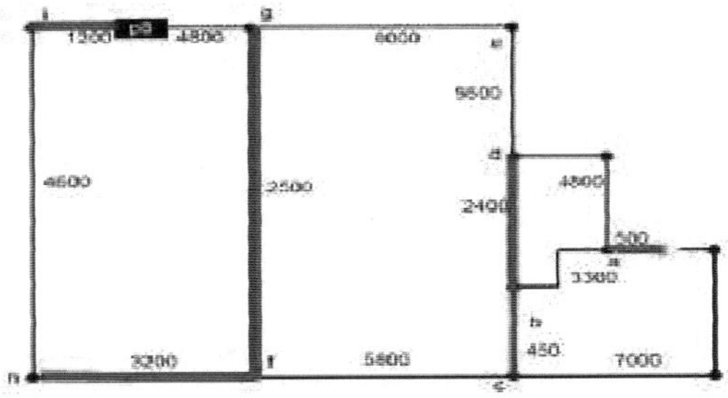

Step 7 (a-b)

Step 8 (h-i)

IOP Conf. Series: Materials Science and Engineering **621** (2019) 012010 doi:10.1088/1757-899X/621/1/012010

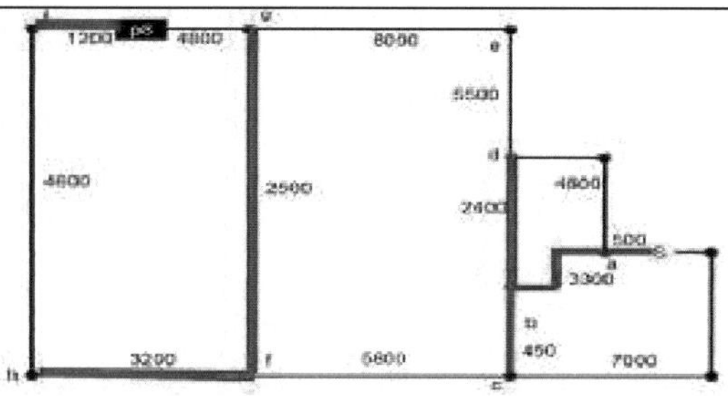

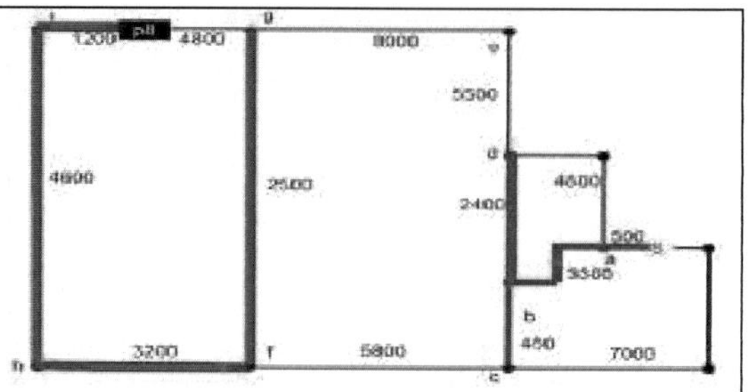

Step 9 (a-d)

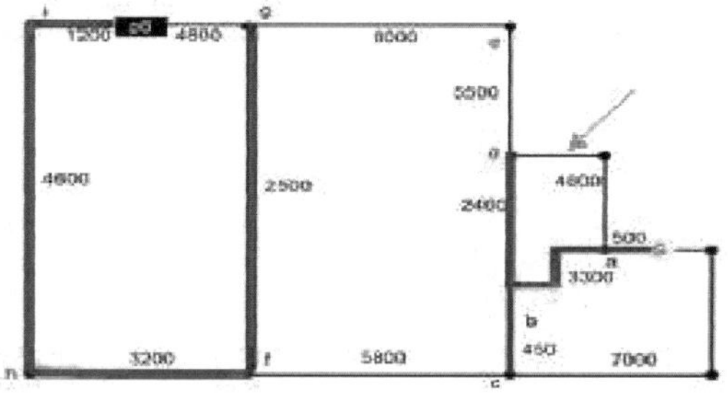

a-d will form a cycle, so it is only depicted with arrows

Step 10 (g-p8)

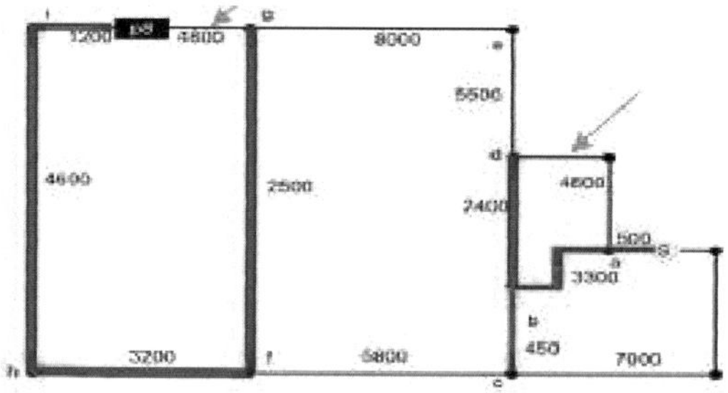

g-p8 will form a cycle, so it is only depicted with arrows

Step 11 (d-e)

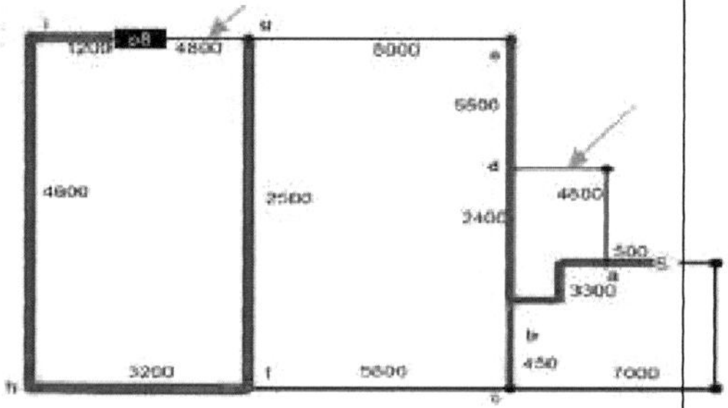

Step 12 (c-f)

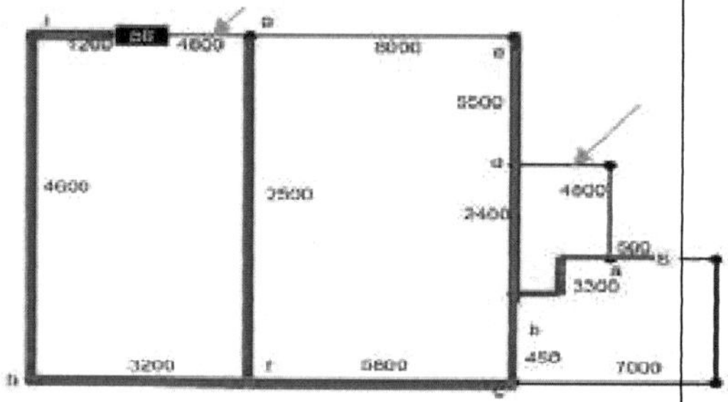

Step 13 (a-c)

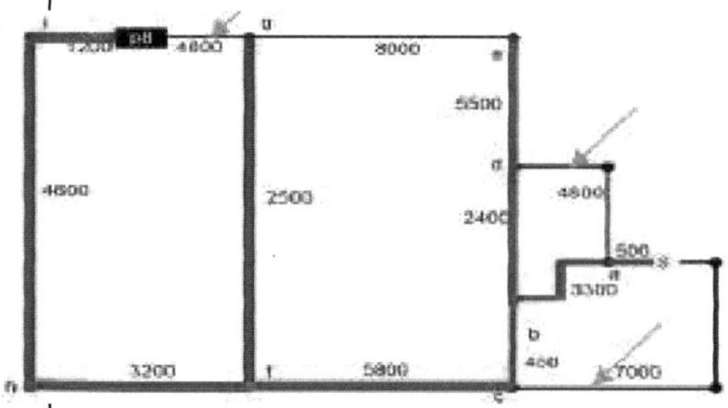

Step 14 (e-g)

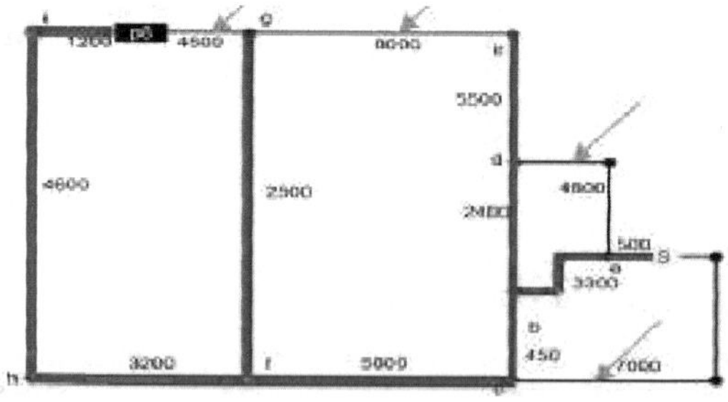

e-g will form a cycle, so it is only depicted with arrow

Figure 6. Stages using the Kruskal algorithm

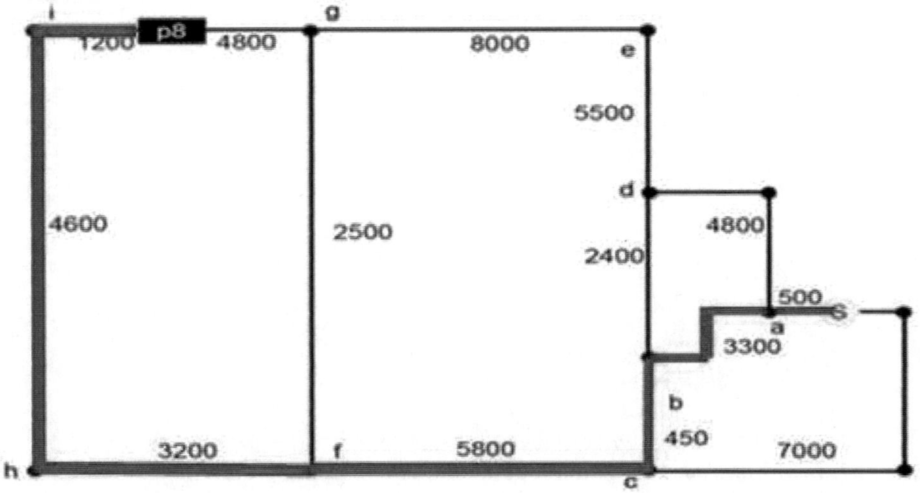

Figure 7. Pathway results using the Kruskal algorithm

***Table 3**. Results of the Kruskal Algorithm*

Ket.	Starting point	Location Destination	Algoritma Kruskal (km)	*Direction API (km)*	Differences
p1	-6.453397, 106.884287	-6.644528, 106.839962	30.3	*29*	*-1.3*
p2	-6.453397, 106.884287	-6.653122, 106.845728	31.5	*31*	-0.5
p3	-6.453397, 106.884287	-6.627004, 106.822351	27.5	*29.1*	1.6
p4	-6.453397, 106.884287	-6.617562, 106.805694	28.5	*30.5*	2
p5	-6.453397, 106.884287	-6.591769, 106.814624	22.95	*24.7*	1.75
p6	-6.453397, 106.884287	-6.591890, 106.793374	22.7	*26.4*	3.7
p7	-6.453397, 106.884287	-6.571830, 106.816978	19.3	*21.9*	2.6
p8	-6.453397, 106.884287	-6.555761, 106.816577	19.0	*20*	1
p9	-6.453397, 106.884287	-6.588110, 106.774094	27.8	*28.2*	0.4
p10	-6.453397, 106.884287	-6.549278, 106.760727	28.6	*31*	2.4
p11	-6.453397, 106.884287	-6.556291, 106.778665	24.2	*27.3*	3.1
p12	-6.453397, 106.884287	-6.536932, 106.773841	25.3	*28*	2.7
			Means		1.6

IOP Conf. Series: Materials Science and Engineering **621** (2019) 012010 doi:10.1088/1757-899X/621/1/012010

4. Conclusion and future research

The results of testing using the Kruskal algorithm and the results of the calculation of google map direction, from s the 12 building stores can be seen in Table 3. From these results it can be concluded that the comparison of the route distance found for the distance of each route, Kruskal algorithm is 83% better than the Google direction API. Thus, a future research are expected to be developed by adding the travel time and traffic conditions at the actual time, and can compare with other shorted path methods.

References

[1] Kusmira M and Rochman T 2017 *Pemanfaatan Aplikasi Graf Pada Pembuatan Jalur Angkot 05 Tasikmalaya, Seminar National Sains dan Teknologi* p-ISSN: 2407 – 1846 e-ISSN: 2460-8416, Fakultas Teknik Universitas Muhammadiyah

[2] Dolfi S et al. 2014 *Algoritma Kruskal Untuk Menentukan Rute Terpendek Pada Jaringan Komputer , Jurnal Ilmiah Mustek* Anim Ha Vol.3 No. 3, ISSN 2089-6697

[3] Hartsfield, Nora R, Gerhard 1990 "Pearls in Graph Theory: A Comprehensive Introduction". United State of America: Academic Press.

[4] Pratama A et al. 2013 *Penggunaan Algoritma Kruskal Dalam Jaringan Pipa Air Minum Kecamatan Nganjuk Kabupaten Nganjuk*, JURNAL SAINS DAN SENI POMITS Vol. 1, No. 1 1-6

[5] Thomas H, Cormen, Charles E, Leiserson, Ronald L. Rivest, and Clifford Stein 2001 Introduction to Algorithms, Second Edition. ISBN 0-262-03293-7. Section 23.2: The algorithms of Kruskal and Prim, pp.567–574. MIT Press and McGraw-Hill.

IOP Conf. Series: Materials Science and Engineering **621** (2019) 012011 doi:10.1088/1757-899X/621/1/012011

Book Embeddings of Infinite Sequences of Extended Periodic Regular Graphs

Benny Pinontoan and Jullia Titaley

Mathematics Department, Sam Ratulangi University, Manado, Indonesia
Corresponding author: bpinonto@unsrat.ac.id

Abstract. A graph is *extended periodic* if it can be made up by gluing identical pieces together with a graph which is isomorphic to or subgraph of that piece, in a circular fashion. We consider an infinite sequence H_n of extended periodic 4-regular graphs and an infinite sequence M_n of extended periodic 5-regular graphs. A *book* is a collection of half-planes called pages, all having a common boundary called the spine. A *book embedding* of a graph G is to put the vertices of G on the spine and each edge of G is contained in one page. The minimum pages needed to embed a graph G into a book is called *pagenumber* of G. We determine the pagenumbers of H_n and M_n to be three.

Keywords: book embedding, pagenumber, extended periodic graph.

1. Introduction

A graph is an ordered pair (N, E) where N is a non-empty set of vertices and E is a set of edges. We consider only simple graphs, i.e. without parallel edges and loops. A graph can be drawn in many ways on the plane. A good drawing has the following properties: no edge crosses itself, no pair of adjacent edges cross, two edges cross at most once, and no more than two edges cross at one point. The *crossing number* $cr(G)$ of a graph G is the minimum number of crossings among all good drawings of G in the plane.

Mohar and Dvořák [5] defined that a graph is *periodic* if it can be obtained by joining identical pieces in a circular fashion. These small pieces are identified as *tiles* by Pinontoan and Richter [6]. We define a graph to be *extended periodic* if it can be made up by joining identical tiles T together with a tile T' which is isomorphic to or subgraph of T, in a circular fashion. Pinontoan and Richter [6] found out that graphs made up by tiles this way, in particular for planar tiles T, then the crossing number of the graph can just be determined by the number of pair-disjoint paths in tile T'. Richter dan Thomassen [7] introduced an infinite sequence H_n ($n \geq 3$) of 4-regular graphs that can be arranged this way. Pinontoan and Richter [6] also studied an infinite sequence M_n ($n \geq 3$) of 5-regular graphs which also can be arranged this way.

A *book* consists of half-planes called *pages*, all having common boundary called the *spine*. A *book embedding* of a graph G is an embedding of G to a book such that the vertices of G lie on the spine and each edge of G lie on a single page. We consider the combinatorial embbeding, i.e, no edge crosses the spine. The *pagenumber* $\rho(G)$ of graph G is the minimal number of pages needed to embed a graph G to a book. Figure 1(a) shows the complete graph K_5 which is already known having crossing number one and Figure 1(b) shows a book embedding of K_5 which needs three pages; this is also the minimal number of pages, and hence $\rho(K_5) = 3$.

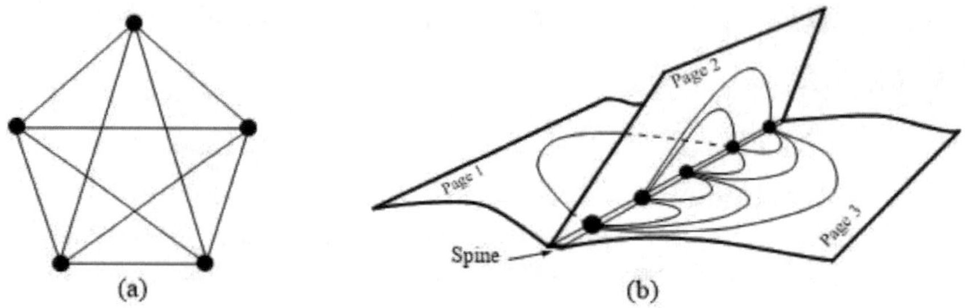

Figure 1. Complete graph K_5 and its book embedding.

Chung, Leighton, and Rosenberg [2] studied pagenumbers of regular graphs and showed that d-regular graphs with n vertices have pagenumber $O(d\sqrt{n})$. Malitz [4] proved that any graph with E edges has pagenumber $O(\sqrt{E})$ and that there is a d-regular N-vertex graph with pagenumber $O(d^{1/2} N^{1/2-1/d})$, where the first bound is tight in the worst case; the second bound is tight when $d > \log N$, and asked if it is tight also for $d < \log n$. Balogh and Salazar [1] showed that it is not so.

In this paper, we determine the pagenumbers of the infinite sequence H_n ($n \geq 3$) of extended periodic 4-regular graphs and the infinite sequence M_n ($n \geq 3$) of extended periodic 5-regular graphs.

2. Infinite Sequences of Extended Periodic 4-regular and 5-regular Graphs

In this section we recall the notion of *tile*, the infinite sequence H_n ($n \geq 3$) of extended periodic 4-regular graphs, and the infinite sequence M_n ($n \geq 3$) of extended periodic 5-regular graphs.

2.1. Tiles

The concept of *tile* was introduced by Pinontoan and Richter [6]. A *tile* $T = (G, L, R)$ is a graph connected G with specified two finite sequences of distinct vertices of G, called the *left-wall* L and *right-wall* R. The *tile crossing number* $tcr(T)$ of $T = (G, L, R)$ is the minimum number of crossings among all drawings of G in the unit square $[0, 1] \times [0, 1]$ on the plane such that the vertices of L appear on $\{0\} \times [0, 1]$ in the order of decreasing y-coordinate and the vertices of R appear on $\{1\} \times [0, 1]$ also in the order of decreasing y-coordinate. A tile T is *planar* if $tcr(T) = 0$. The tile $H = ((\{a, b, c, d, e\}\ \{ac, ad, cb, be, ce, cd\}), (a, b), (d, e))$ in Figure 2(a) is planar whereas the tile $H' = ((\{a, b, c, d, e\}\ \{ac, ad, cb, be, ce, cd\}), (a, b), (e, d))$ in Figure 2(b) is not planar.

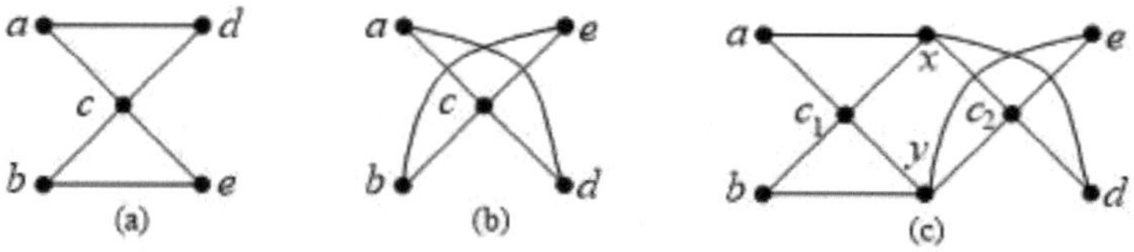

Figure 2. Tiles H, H', and HH'.

Tile $T_1 = (G_1, L_1, R_1)$ is compatible with tile $T_2 = (G_2, L_2, R_2)$ if $|R_1| = |L_2|$. If $T_1 = (G_1, L_1, R_1)$ is compatible with $T_2 = (G_2, L_2, R_2)$, then R_1 and L_2 can be glued together, i.e. by identifying them, to get a bigger tile $T_1T_2 = (G_1 \cup G_2, L_1, R_2)$. Figure 2(c) shows tile HH' as result of gluing tile H in Figure 2(a) and tile H' in Figure 2(b), where the right wall-of H is identified with the left-wall of H', that is $d = a$ and $e = b$, then rename the inner vertices.

A tile is self-*compatible* if it is compatible to itself. If T is a *self-compatible* tile, then T^n is obtained by gluing n copies of T in a linear fashion and $o(T^n)$ is obtained by gluing the first and the last copies of T in T^n together, to get a circular fashion of arrangement. Note that T^n is a tile and $o(T^n)$ is a graph. The *twist* T' of $T = (G, L, R)$ is the tile obtained by reversing the order of R. The tile H' is Figure 2(b) is the twist of the tile H in Figure 2(a). If T and S are compatible tiles, and T is a self-compatible tile, then we define $T^{\otimes}(n) = o(T^nT')$ where T' is the twist of T, and $T(n, S) = o(T^nS)$.

2.2. Infinite sequence H_n ($n \geq 3$) of extended periodic 4-regular graphs

Richter dan Thomassen [7] introduced the infinite sequence H_n ($n \geq 3$) of 4-regular graphs which were obtained from a cycle C_{2m} of length $2m$ by adding, and for each two diametrically oppose edges e and e' on C_{2m}, a new vertex joined to the four ends of the edges e and e'. The vertices of H_n are $a_1, a_2, a_3, ..., a_{n+1}, b_1, b_2, b_3, ..., b_{n+1}, c_1, c_2, c_3, ...,$ and c_{n+1}. The edges of H_1 are a_kc_k, b_kc_k ($1 \leq k \leq n+1$), a_ka_{k+1}, b_kb_{k+1}, c_ka_{k+1}, c_kb_{k+1} ($1 \leq k \leq n$), a_1b_{n+1}, a_1c_{n+1}, b_1a_{n+1}, and b_1c_{n+1}.

Pinontoan and Richter [6] identified H_n as a graph made up by gluing n planar tiles H in Figure 2(a) and a single tile H' in Figure 2(b) which is the twist of H and isomorphic to H, in circular fashion, that is, $H_n = H^{\otimes}(n) = o(H^nH')$, and rename the vertices with appropiate names. Figure 3 shows B_5.

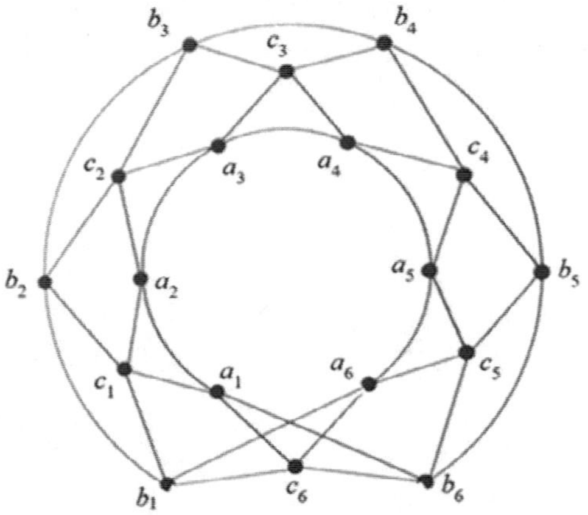

Figure 3. Graph H_5.

2.3. Infinite sequence M_n ($n \geq 3$) of extended periodic 5-regular graphs

Let $n \geq 3$. The graph M_n is the 5-regular graph consisting of $4n + 2$ vertices $a_1, b_1, c_1, d_1, a_2, b_2, c_2, d_2, ..., a_n, b_n, c_n, d_n, e,$ and f, and containing the cycle $a_1c_1a_2c_2 ... a_nc_ne$, and also the edges a_kb_k, c_kd_k, a_kd_k, b_kc_k ($1 \leq k \leq n$), the edges c_kb_{k+1}, d_ka_{k+1}, b_kb_{k+1}, d_kd_{k+1} ($1 \leq k \leq n-1$), the edges eb_1, ef, d_ne, b_nf, c_nf, fa_1, fd_1, and d_nb_1. Figure 4 shows M_3.

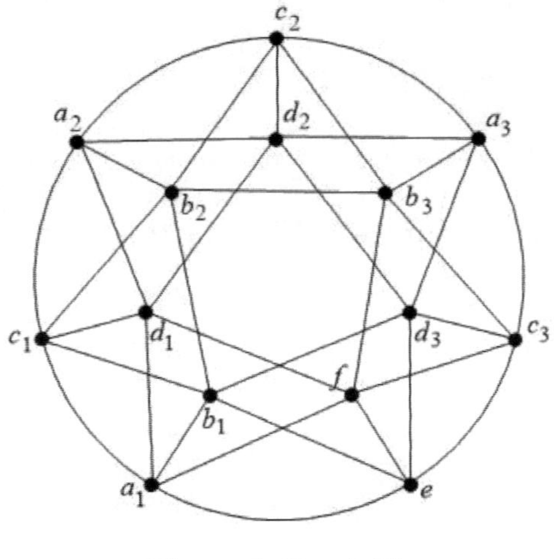

Figure 4. Graph M_3.

The graph M_n ($n \geq 3$) can also be made up by gluing n identical planar tiles M pictured in Figure 5(a) and one tile S in Figure 5(b) which is a subgraph of M, in circular fashion, that is, $M_n = M(n, S) = \text{o}(M^n S)$, and then contract the vertices of degree two.

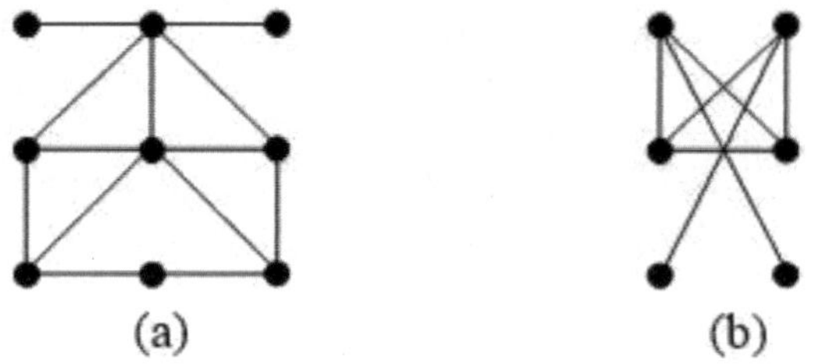

Figure 5. Tile M and tile S.

Figure 6 shows $M_3 = M(3, S) = \text{o}(M^3 S)$, that is obtained by gluing three tiles M in Figure 5(a) and one tile S in Figure 5(b) in a circular fashion.

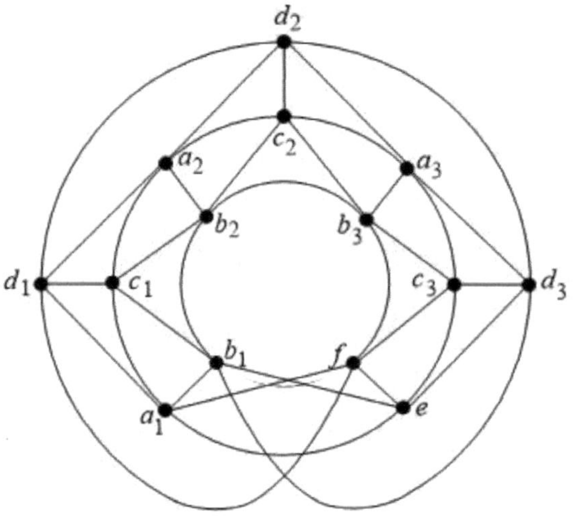

Figure 6. Graph M_3.

3. Book Embeddings of Hn and Mn

In this section we show that $\rho(H_n) = 3$ and $\rho(M_n) = 3$, for $n \geq 3$.

Theorem 1. Let $n \geq 3$. Then $\rho(H_n) = 3$.

Proof. Richter dan Thomassen [7] determined that the crossing number of H_n ($n \geq 3$) is 3, which was established by Pinontoan and Richter [6] using tile mechanism. So M_n is not planar on the plane (two pages), it needs more than two pages to embed and hence $\rho(H_n) \geq 3$.

We show that $\rho(H_n) \leq 3$ by drawing an embedding to book. The graph H_n is embbeded to a book as follows. Put the vertices on the spine in the following order:

$$a_1, b_1, c_1, a_2, b_2, c_2, \ldots, a_{n+1}, b_{n+1}, c_{n+1}.$$

Then on the first page, put the edges $b_i c_i$ ($1 \leq i \leq n + 1$), $b_i b_{i+1}$ ($1 \leq i \leq n$), $c_i b_{i+1}$ ($1 \leq i \leq n$), and $a_1 b_{n+1}$ dan $a_1 c_{n+1}$ without crossing. On the second page, put the edges $a_i c_i$ ($1 \leq i \leq n + 1$), $a_i a_{i+1}$ ($1 \leq i \leq n$), and $c_i a_{i+1}$ ($1 \leq i \leq n$) without crossing, and on the third page, put the edges $b_1 a_{n+1}$ and $b_1 c_{n+1}$ without crossing. So we need at least three pages to put the edges and hence $\rho(H_n) \leq 3$. Therefore $\rho(H_n) = 3$. □

IOP Conf. Series: Materials Science and Engineering **621** (2019) 012011 doi:10.1088/1757-899X/621/1/012011

Figure 7 shows a book embedding of H_5. The vertices are lined up on the spine, the edges drawn above the vertices are on page 1, the continous edges under the vertices are on page 2, and the edges drawn with stipple lines under the vertices are on page 3.

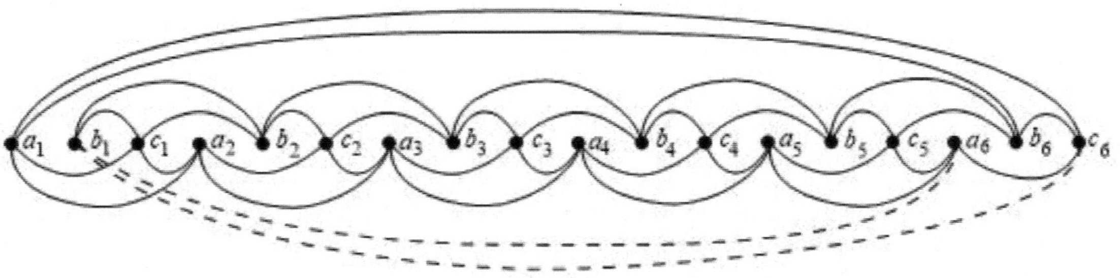

Figure 7. A book embedding of H_5.

Theorem 2. Let $n \geq 3$. Then $\rho(M_n) = 3$.

Proof. Pinontoan and Richter [6] have determined that the crossing numbers of M_n ($n \geq 3$) is 6. M_n is not planar and so it needs at least three pages. Hence $\rho(M_n) \geq 3$.

To show that $\rho(M_n) \leq 3$ we draw an embedding of M_n to book on three pages as follows. Put the vertices on the spine in the following order a_1, b_1, c_1, d_1, a_2, b_2, c_2, d_2, ... , a_n, b_n, c_n, d_n, e, and f. Then put the following edges on the first page: $a_k b_k$, $b_k c_k$, $c_k d_k$ $(1 \leq i \leq n)$, $b_k b_{k+1}$ $(1 \leq i \leq n-1)$, $b_k b_{k+1}$ $(1 \leq i \leq n-1)$, and the edges $c_n e$, $d_n e$, $e f$, $b_n f$, and $c_n f$. On the second page, put the edges $a_k c_k$, $a_k d_k$ $(1 \leq i \leq n)$, $d_k d_{k+1}$ $(1 \leq i \leq n-1)$, and the edge $d_1 f$. Finally, on the third page, put the edges $c_k b_{k+1}$ $(1 \leq i \leq n-1)$, and also the edges $a_1 e$, $a_1 f$, $b_1 e$, and $b_1 d_n$. So we need at least three pages to put the edges of M_n and hence $\rho(M_n) \leq 3$. Therefore $\rho(M_n) = 3$. □

A book embedding of M_3 can be seen in Figure 8, where the edges shown in Figure 8(a) above the vertices are on the first page and the edges under the vertices are on the second page. Finally, the edges in Figure 8(b) are put on the third page.

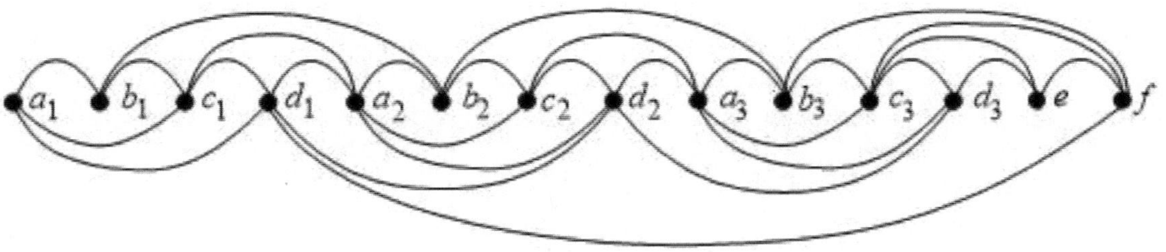

(a) Page 1 and Page 2

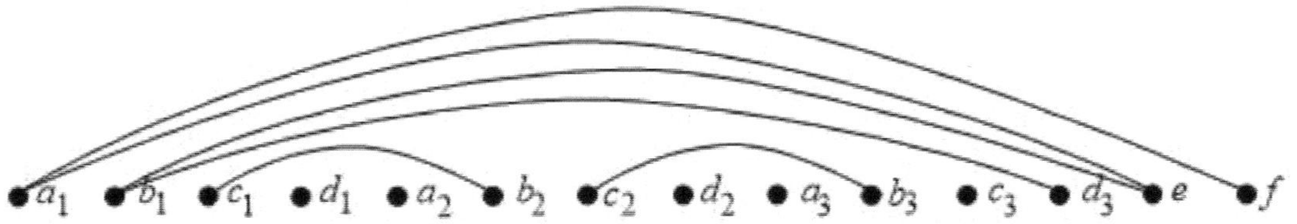

(b) Page 3

Figure 8. A book embedding of M_3.

4. Acknowledgments

This research was funded by Sam Ratulangi University Grant DIPA No. SP DIPA – 042.01.2400959/2017 dated 21 April 2017.

References

[1] Balogh J and Salazar G 2015 Book embeddings of regular graphs. *SIAM Journal in Discrete Mathematics* **29** 811-22.

[2] Chung F R T , Leighton F T and Rosenberg A L 1987 Embedding graphs in books: a layout problem with applications to VLSI design, *SIAM Journal on Algebraic and Discrete Methods* **8** 33-58.

[3] Kainen P C 1974 An introduction to topological graph theory, in *Graphs and Combinatorics*, Ed R Bari and F Harary (New York: Springer) pp 76-108.

[4] Malitz S M 1994 Graphs with E edges have pagenumber O(√E), *J. Algorithms* **17** (1) 71–84.

[5] Mohar B and Dvořák Z 2016 Crossing numbers of periodic graphs. *J. Graph Theory* **83** 34-43.

[6] Pinontoan B and Richter R B 2003 Crossing numbers of sequences of graphs II: planar tiles. *J. Graph Theory* **42** 332-41.

[7] Richter R B and Thomassen C 1993 Minimal graphs with crossing number at least *k*. *J. Combin. Theory Ser. B.* **58** (2) 217-23.

IOP Conf. Series: Materials Science and Engineering **621** (2019) 012012 doi:10.1088/1757-899X/621/1/012012

A new 4-D hyperchaotic hyperjerk system with a single equilibrium, its dynamic properties and circuit design

S Vaidyanathan[1], A Sambas[2], Sukono[3], M Mamat[4], W S Mada Sanjaya[5]

[1]Research and Development Centre, Vel Tech University, Chennai, Tamil Nadu, INDIA
[2]Department of Mechanical Engineering, Universitas Muhammadiyah Tasikmalaya, INDONESIA
[3]Department of Mathematics, Faculty of Mathematics and Natural Sciences Universitas Padjadjaran, INDONESIA
[4]Faculty of Informatics and Computing, Universiti Sultan Zainal Abidin, KualaTerengganu, MALAYSIA
[5]Department of Physics, Universitas Islam Negeri Sunan Gunung Djati Bandung, INDONESIA

Corresponding author's e-mail address: sundarvtu@gmail.com

Abstract. This paper announces a new four-dimensional hyperchaotic hyperjerk system with a single equilibrium and discusses its dynamic properties such as Lyapunov exponents, phase portraits, Kaplan-Yorke dimension and equilibrium points. Hyperjerk systems have a nice triangular structure in their dynamics and they have many engineering applications. Our new hyperjerk system has three nonlinearities in total. New synchronization results based on active backstepping control are also derived for the new hyperjerk system. In addition, an electronic circuit implementation of the new hyperjerk system is designed carefully and examined well in MultiSIM. A good qualitative agreement has been shown between the MATLAB simulations of the theoretical hyperchaotic hyperjerk model and the MultiSIM results.

1. Introduction

Chaotic dynamical systems have applications in several areas of science and engineering [1-4]. Some popular applications of chaotic systems can be cited as oscillators [5-10], weather systems [11-12], chemical reactors [13-15], robotics [16-17], communication systems [18-20], circuits [21-26], etc.

In 1996, Gottlieb [27] exhibited that 3-D chaotic systems can be expressed as single ordinary differential equations that can be termed as jerk differential equations. The jerk differential equations arise in many physical models of science such as jerk circuits [28-30], thermal arc plasma [31], biological reactions [32], mechanical oscillations [33-34], etc.

In physics, a jerk differential equation can be expressed as the third order dynamics

$$D^3x = f(x, Dx, D^2x) \tag{1}$$

where $D = \dfrac{d}{dt}$. If $x(t)$ represents the *displacement* of a body, then $Dx = \dot{x}$ represents its *velocity* and $D^2x = \ddot{x}$ its *acceleration*. The third derivative $D^3x = \dddot{x}$, is called as the *jerk* of the body.

Generalizing the jerk system (1), we obtain the hyperjerk system given by the dynamics

$$D^{(n)}x = f(x, Dx, \ldots, D^{(n-1)}x), \quad (n \geq 4) \tag{2}$$

It is convenient to express the hyperjerk dynamics (3) in the form of a system of ordinary differential equations. For this purpose, we define phase variables as

$$x_1 = x,\ x_2 = Dx,\ \ldots,\ x_n = D^{(n-1)}x \tag{3}$$

Using the phase variables (3), we can rewrite the hyperjerk dynamics (2) in system form as

$$\begin{cases} \dot{x}_1 = x_2 \\ \dot{x}_2 = x_3 \\ \vdots \quad \vdots \quad \vdots \\ \dot{x}_{n-1} = x_n \\ \dot{x}_n = f(x_1, x_2, \ldots, x_n) \end{cases} \tag{4}$$

Hyperjerk systems have generated significant attention in the chaos literature in view of their simple structure and complex dynamic properties [35-40].

Hyperchaotic systems are chaotic systems with more than one positive Lyapunov exponent, which have high level of complexity [41-45].

In this paper, we modify the Daltzis hyperjerk system [46] to obtain a new hyperjerk system. It is interesting that our new hyperjerk system is hyperchaotic and has a unique equilibrium at the origin. We detail our new system with phase portraits, an analysis of its Lyapunov exponents, Kaplan-Yorke dimension, dissipativity, etc.

As a control application, we describe the synchronization of the new hyperjerk system with backstepping control [47-48]. As an engineering application, we design an electronic circuit of the new hyperjerk system using MultiSIM.

The structure of this paper is as follows. Section 2 announces the new hyperchaotic hyperjerk system with a unique equilibrium and describes its properties. Section 3 details the synchronization of the new hyperjerk system via backstepping control. Section 4 describes an electronic circuit design of the new hyperjerk system using MultiSIM. Section 5 summarizes the paper with conclusions.

2. A new 4-D hyperchaotic hyperjerk system

In 2018, Daltzis et al. [46] proposed a new 4-D hyperjerk system given by the dynamics

$$\begin{cases} \dot{x}_1 = x_2 \\ \dot{x}_2 = x_3 \\ \dot{x}_3 = x_4 \\ \dot{x}_4 = -x_1 - x_2 - ax_3 - b\,|\,x_2\,| - cx_1^4 x_4 \end{cases} \tag{5}$$

In Eq. (5), x_1, x_2, x_3, x_4 are the states and a, b, c are positive parameters. In [49], it was shown that the hyperjerk system (5) is hyperchaotic for the choice of parameters $(a, b, c) = (3.8, 0.1, 1.5)$.

Using Wolf's algorithm [49], the Lyapunov exponents of the Daltzis hyperjerk system (5) are calculated for $(a, b, c) = (3.8, 0.1, 1.5)$, and $X(0) = (0.1, 0.1, 0.1, 0.1)$ for $T = 1E4$ seconds as

$$L_1 = 0.1201,\ L_2 = 0.0210,\ L_3 = 0,\ L_4 = -1.2854 \tag{6}$$

Since the Lyapunov exponents L_1 and L_2 are positive in (6), we conclude that the Daltzis hyperjerk system (5) is hyperchaotic. Also, we can calculate the Kaplan-Yorke dimension of the Daltzis hyperjerk system (5) as follows:

$$D_{KY} = 3 + \frac{L_1 + L_2 + L_3}{|\,L_4\,|} = 3.1098 \tag{7}$$

In this paper, we propose a new 4-D hyperchaotic hyperjerk system by modifying the Daltzis hyperjerk system (5) and obtaining a new system given as follows:

$$\begin{cases} \dot{x}_1 = x_2 \\ \dot{x}_2 = x_3 \\ \dot{x}_3 = x_4 \\ \dot{x}_4 = -x_1 - x_2 - ax_3 + b\,|\,x_3\,| - cx_1^4 x_4 - dx_2^2 \end{cases} \tag{8}$$

In Eq. (8), x_1, x_2, x_3, x_4 are the states and a, b, c, d are positive parameters.

We show that the system (8) is hyperchaotic for the parameter values

$$(a, b, c, d) = (3.6, 0.02, 3, 0.05) \tag{9}$$

Using Wolf's algorithm [49], the Lyapunov exponents of the new hyperjerk system (8) are obtained for the parameter values as in (9) and $X(0) = (0.1, 0.1, 0.1, 0.1)$ for $T = 1E4$ seconds as

$$L_1 = 0.12479, \; L_2 = 0.03995, \; L_3 = 0, \; L_4 = -1.22715 \tag{10}$$

Since L_1 and L_2 are positive Lyapunov exponents, the new hyperjerk system (8) is hyperchaotic.

By adding all the Lyapunov exponents in (10), we get their sum as

$$L_1 + L_2 + L_3 + L_4 = -1.06241 < 0 \tag{11}$$

This shows that the new hyperchaotic hyperjerk system (8) is dissipative.

Comparing the equations (6) and (10), we notice that the positive Lyapunov exponents L_1, L_2 of the new hyperjerk system (8) are greater than the positive Lyapunov exponents L_1, L_2 of the Daltzis hyperjerk system (5).

The Kaplan-Yorke fractal dimension of the new hyperjerk system (8) is calculated as

$$D_{KY} = 3 + \frac{L_1 + L_2 + L_3}{|\,L_4\,|} = 3.1342, \tag{12}$$

which is greater than the Kaplan-Yorke dimension D_{KY} of the Daltzis hyperjerk system (5).

Thus, we have shown that the new hyperjerk system (8) has more complex dynamic properties than the Daltzis hyperjerk system (5).

It is easy to verify that $E_0 = (0, 0, 0, 0)$ is the unique equilibrium point of the new hyperjerk system (8). The Jacobian matrix of the new hyperjerk system (8) at E_0 is calculated for the hyperchaotic case (9) as

$$J_0 = \begin{bmatrix} 0 & 1 & 0 & 0 \\ 0 & 0 & 1 & 0 \\ 0 & 0 & 0 & 1 \\ -1 & -1 & -3.6 & 0 \end{bmatrix} \tag{13}$$

The eigenvalues of J_0 are computed as

$$\lambda_{1,2} = 0.1604 \pm 1.8395i, \; \lambda_{3,4} = -0.1604 \pm 0.5172i \tag{14}$$

This shows that the equilibrium point $E_0 = (0, 0, 0, 0)$ is in the critical case.

Figures 1-4 show the 2-D phase portraits of the new hyperjerk system (8).

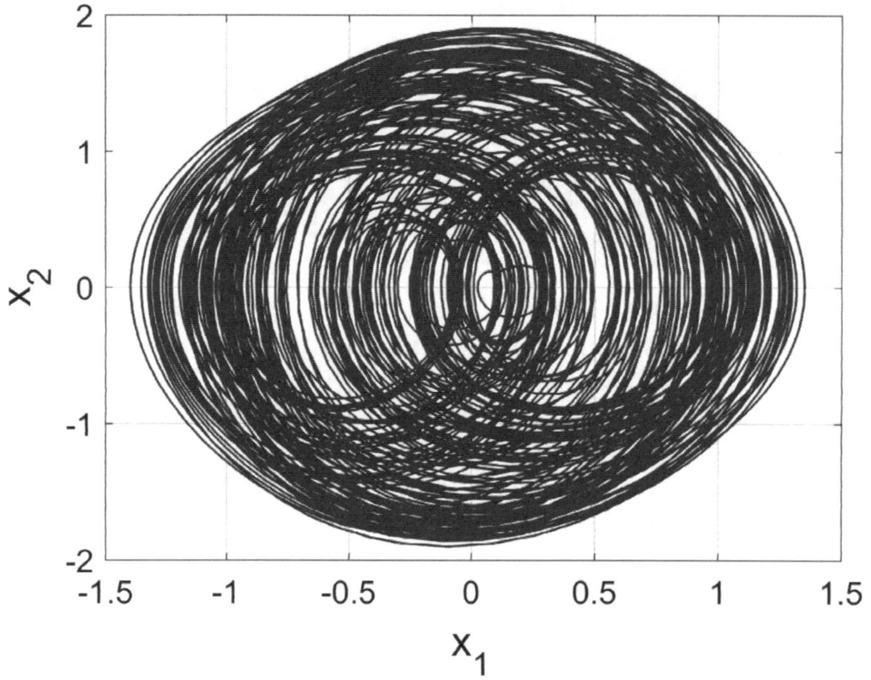

Figure 1. Two-dimensional phase plot of the new hyperjerk system (8) in the $(x_1, x_2)-$ plane with $X_0 = (0.1, 0.1, 0.1, 0.1)$ and $(a,b,c,d) = (3.6, 0.02, 3, 0.05)$

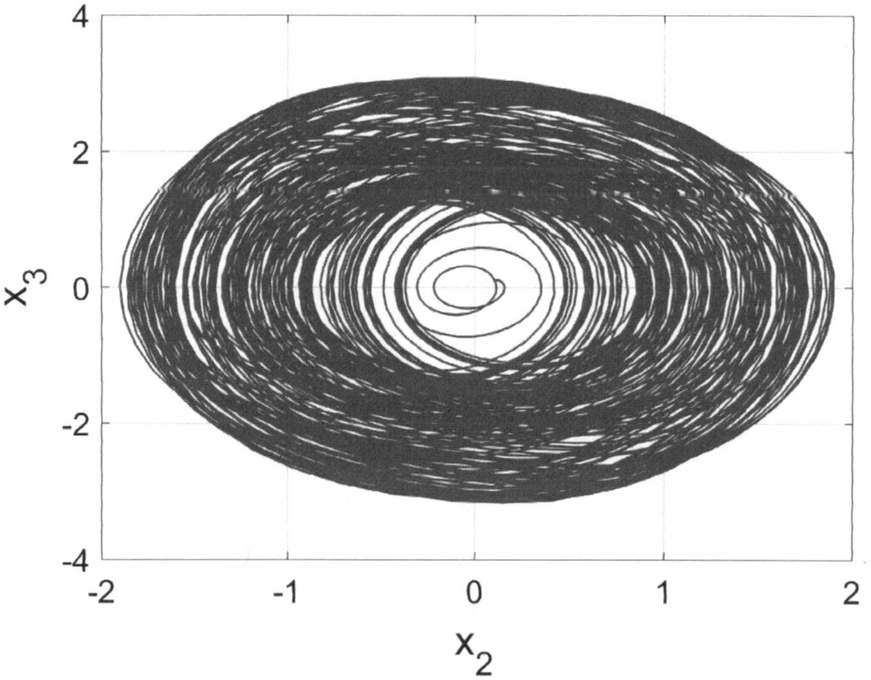

Figure 2. Two-dimensional phase plot of the new hyperjerk system (8) in the $(x_2, x_3)-$ plane with $X_0 = (0.1, 0.1, 0.1, 0.1)$ and $(a,b,c,d) = (3.6, 0.02, 3, 0.05)$

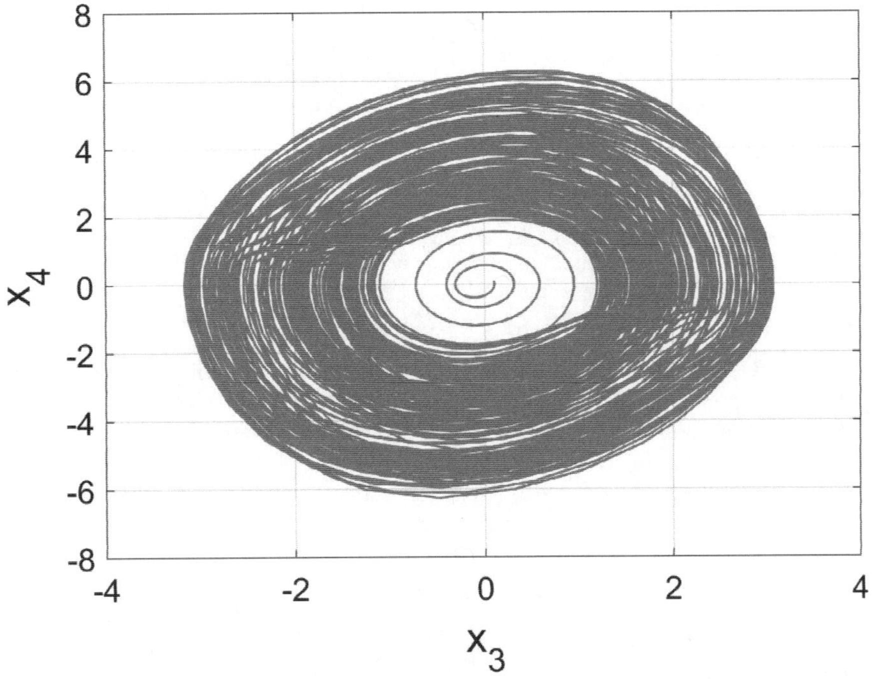

Figure 3. Two-dimensional phase plot of the new hyperjerk system (8) in the $(x_3, x_4)-$ plane with $X_0 = (0.1, 0.1, 0.1, 0.1)$ and $(a, b, c, d) = (3.6, 0.02, 3, 0.05)$

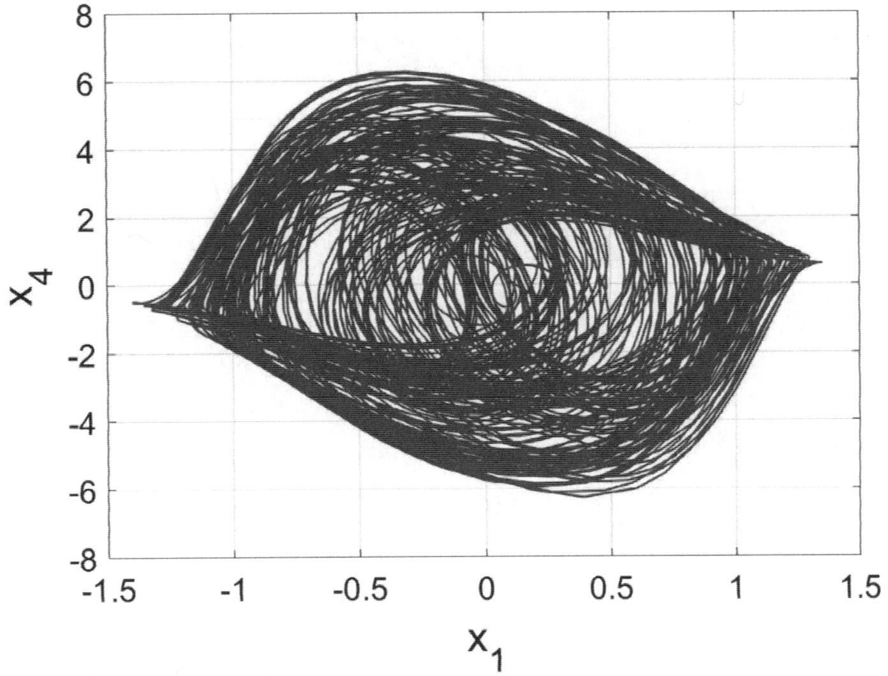

Figure 4. Two-dimensional phase plot of the new hyperjerk system (8) in the $(x_1, x_4)-$ plane with $X_0 = (0.1, 0.1, 0.1, 0.1)$ and $(a, b, c, d) = (3.6, 0.02, 3, 0.05)$

3. Global hyperchaos synchronization of the new hyperjerk systems via backstepping control

In this section, we design an engineering application of the new hyperjerk system, viz. global hyperchaos synchronization of the new hyperjerk systems taken as master and slave systems via active backstepping control method.

As the master system, we take the new hyperjerk system given by

$$\begin{cases} \dot{x}_1 = x_2 \\ \dot{x}_2 = x_3 \\ \dot{x}_3 = x_4 \\ \dot{x}_4 = -x_1 - x_2 - ax_3 + b\,|\,x_3\,| - cx_1^4 x_4 - dx_2^2 \end{cases} \tag{15}$$

where $X = (x_1, x_2, x_3, x_4)$ is the state and a, b, c, d are system parameters.

As the slave system, we consider the new hyperjerk system with control given by

$$\begin{cases} \dot{y}_1 = y_2 \\ \dot{y}_2 = y_3 \\ \dot{y}_3 = y_4 \\ \dot{y}_4 = -y_1 - y_2 - ay_3 + b\,|\,y_3\,| - cy_1^4 y_4 - dy_2^2 + u \end{cases} \tag{16}$$

where $Y = (y_1, y_2, y_3, y_4)$ is the state and u is the active backstepping control to be designed.

The synchronization error between new hyperjerk systems (15) and (16) is defined via the equation

$$\begin{cases} e_1 = y_1 - x_1 \\ e_2 = y_2 - x_2 \\ e_3 = y_3 - x_3 \\ e_4 = y_4 - x_4 \end{cases} \tag{17}$$

The dynamics of the synchronization error is easily obtained as follows:

$$\begin{cases} \dot{e}_1 = e_2 \\ \dot{e}_2 = e_3 \\ \dot{e}_3 = e_4 \\ \dot{e}_4 = -e_1 - e_2 - ae_3 + b(|\,y_3\,| - |\,x_3\,|) - c(y_1^4 y_4 - x_1^4 x_4) - d(y_2^2 - x_2^2) + u \end{cases} \tag{18}$$

Using active backstepping control method, we establish the key result of this section.

Theorem 1. The master and slave hyperchaotic systems represented by the new hyperjerk systems (15) and (16) are globally and exponentially synchronized by means of the active backstepping feedback control law given by

$$u = -4e_1 - 9e_2 - (9-a)e_3 - 4e_4 - b\left(|\,y_3\,| - |\,x_3\,|\right) + c\left(y_1^4 y_4 - x_1^4 x_4\right) + d\left(y_2^2 - x_2^2\right) - k\xi_4 \tag{19}$$

where $k > 0$ is a gain constant and

$$\xi_4 = 3e_1 + 5e_2 + 3e_3 + e_4 \tag{20}$$

Proof. We establish this result via backstepping control method and Lyapunov stability theory [50]. We define the Lyapunov function

$$V_1(\xi_1) = \frac{1}{2}\xi_1^2 \tag{21}$$

where

$$\xi_1 = e_1 \tag{22}$$

Differentiating V_1 along the error dynamics (18), we get

$$\dot{V}_1 = \xi_1\dot{\xi}_1 = e_1 e_2 = -\xi_1^2 + \xi_1(e_1 + e_2) \tag{23}$$

We set

$$\xi_2 = e_1 + e_2 \tag{24}$$

Using (24), we can rewrite (23) as

$$\dot{V}_1 = -\xi_1^2 + \xi_1\xi_2 \qquad (24)$$

Next, we define the Lyapunov function

$$V_2(\xi_1, \xi_2) = V_1(\xi_1) + \frac{1}{2}\xi_2^2 = \frac{1}{2}(\xi_1^2 + \xi_2^2) \tag{26}$$

Differentiating V_2 along the error dynamics (18), we get

$$\dot{V}_2 = -\xi_1^2 - \xi_2^2 + \xi_2(2e_1 + 2e_2 + e_3) \tag{27}$$

Now, we set

$$\xi_3 = 2e_1 + 2e_2 + e_3 \tag{28}$$

Using (28), we can simplify Eq. (27) as

$$\dot{V}_2 = -\xi_1^2 - \xi_2^2 + \xi_2\xi_3 \tag{29}$$

Next, we define the Lyapunov function

$$V_3(\xi_1, \xi_2, \xi_3) = V_2(\xi_1, \xi_2) + \frac{1}{2}\xi_3^2 = \frac{1}{2}(\xi_1^2 + \xi_2^2 + \xi_3^2) \tag{30}$$

Differentiating V_3 along the error dynamics (18), we get

$$\dot{V}_3 = -\xi_1^2 - \xi_2^2 - \xi_3^2 + \xi_3(3e_1 + 5e_2 + 3e_3 + e_4) \tag{31}$$

Next, we set

$$\xi_4 = 3e_1 + 5e_2 + 3e_3 + e_4 \tag{32}$$

Using (32), we can simplify the equation (31) as

$$\dot{V}_3 = -\xi_1^2 - \xi_2^2 - \xi_3^2 + \xi_3\xi_4 \tag{33}$$

To simplify the notation, we set $\xi = (\xi_1, \xi_2, \xi_3, \xi_4)$.

Finally, we define the quadratic Lyapunov function

$$V(\xi) = V_3(\xi_1, \xi_2, \xi_3) + \frac{1}{2}\xi_4^2 = \frac{1}{2}\left(\xi_1^2 + \xi_2^2 + \xi_3^2 + \xi_4^2\right) \tag{34}$$

Clearly, V is a quadratic and positive definite function on R^4.

Differentiating V along the error dynamics (18), we get

$$\dot{V} = -\xi_1^2 - \xi_2^2 - \xi_3^2 - \xi_4^2 + \xi_4(\xi_4 + \xi_3 + \dot{\xi}_4) = -\xi_1^2 - \xi_2^2 - \xi_3^2 - \xi_4^2 + \xi_4(S) \tag{35}$$

where

$$S = \xi_4 + \xi_3 + \dot{\xi}_4 = \xi_4 + \xi_3 + (3\dot{e}_1 + 5\dot{e}_2 + 3\dot{e}_3 + \dot{e}_4) \tag{36}$$

It is easy to see that

$$S = 4e_1 + 9e_2 + (9-a)e_3 + 4e_4 + b(|\,y_3\,| - |\,x_3\,|) - c(y_1^4 y_4 - x_1^4 x_4) - d(y_2^2 - x_2^2) + u \tag{37}$$

Substituting the backstepping control law (19) into (37), we get

$$S = -k\xi_4 \tag{38}$$

Using (38), we can simplify Eq. (35) as

$$\dot{V} = -\xi_1^2 - \xi_2^2 - \xi_3^2 - (1+k)\xi_4^2 \tag{39}$$

this is a quadratic and negative definite function on R^4.

Thus, the proof is complete by Lyapunov stability theory. ■

For numerical simulations, we take the parameters of the hyperjerk systems as in the hyperchaotic case, *viz.* $(a,b,c,d)=(3.6,0.02,3,0.05)$. Also, we take $k=25$.

The initial conditions of the master system (15) are taken as $X(0)=(3.6,1.7,0.1,1.4)$.

The initial conditions of the slave system (16) are taken as $Y(0)=(2.1,0.4,1.2,2.5)$.

Figure 5 shows the complete synchronization of the new hyperjerk systems represented by the master system (15) and the slave system (16), while Figure 6 depicts the time-history of the synchronization error between the systems (15) and (16).

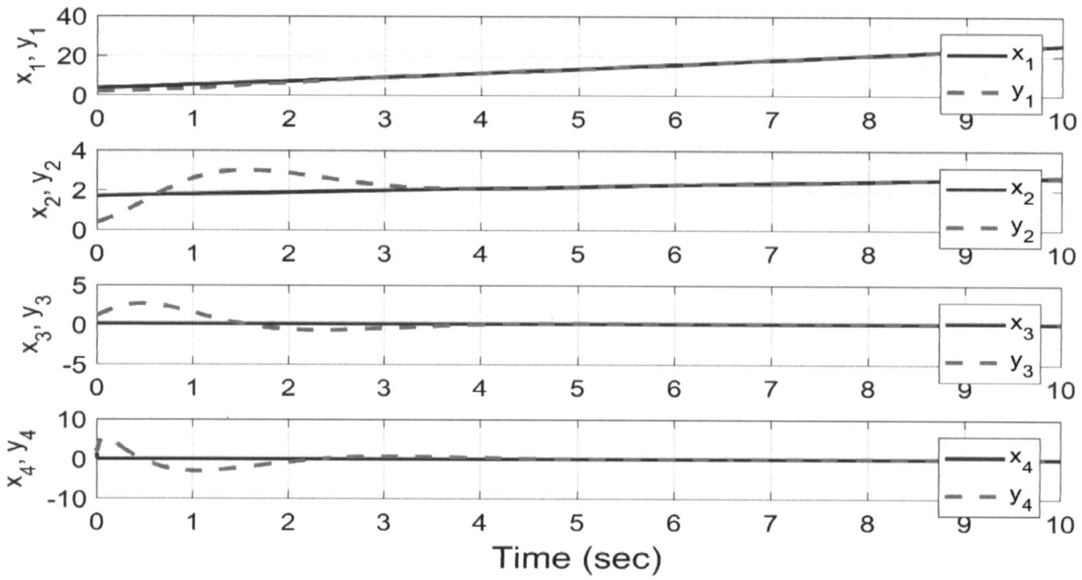

Figure 5. Complete synchronization of the hyperjerk systems (15) and (16) with $X(0)=(3.6,1.7,0.1,1.4)$ **and** $Y(0)=(2.1,0.4,1.2,2.5)$

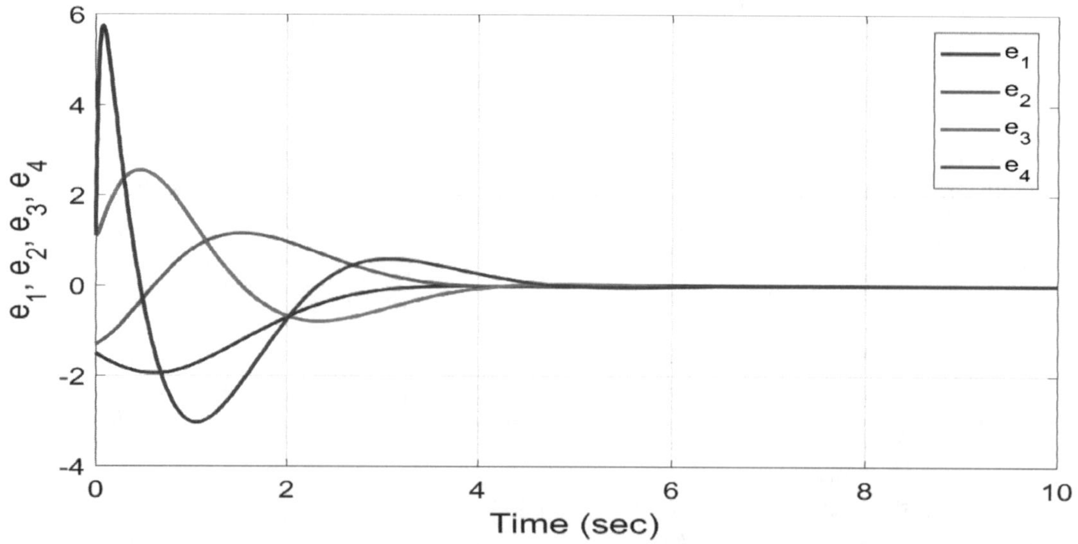

Figure 6. Time-history of the synchronization error between the new hyperjerk systems (15) and (16) with $X(0)=(3.6,1.7,0.1,1.4)$ **and** $Y(0)=(2.1,0.4,1.2,2.5)$

4. Circuit realization of the new hyperjerk system

The new chaotic system described by (1) is easily implemented with an electronic circuit via op-amps and multipliers [51-54], as shown in Figure 7. As shown in this figure, the circuit includes twenty two resistors, four capacitors, ten operational amplifiers (TL082CD), four analog multipliers (AD633JN) and two diodes (1N4148).

Chaotic differential equations of the new circuit are given below.

$$\left.\begin{aligned} \dot{x}_1 &= \frac{1}{C_1R_1}x_2 \\ \dot{x}_2 &= \frac{1}{C_2R_2}x_3 \\ \dot{x}_3 &= \frac{1}{C_3R_3}x_4 \\ \dot{x}_4 &= -\frac{1}{C_4R_4}x_1 - \frac{1}{C_4R_5}x_2 - \frac{1}{C_4R_6}x_3 + \frac{1}{C_4R_7}|x_3| - \frac{1}{100C_4R_8}x_1^4x_4 - \frac{1}{C_4R_9}x_2^2 \end{aligned}\right\} \quad (40)$$

In which the voltages of capacitors are denoted as x_1, x_2, x_3, x_4. When the system parameters are $(a,b,c,d)=(3.6,0.02,3,0.05)$, the circuit elements have the values R_6 = 27.77 kΩ, R_7 = 5 MΩ, R_8 = 333.33 Ω, R_9 = 2 MΩ, $R_1 = R_2 = R_3 = R_4 = R_5 = R_{10} = R_{11} = R_{12} = R_{13} = R_{14} = R_{15} = R_{16} = R_{17} = R_{18} = R_{19} = R_{20} = R_{21} = R_{22}$ =100 kΩ, $C_1 = C_2 = C_3 = C_4$ = 3.2 nF. Phase portrait outputs of the electronic circuit simulation are shown in Figure 8. Circuit simulation outputs (see Figure 8) show good qualitative matching with the numerical simulations (see Figures 1-4).

5. Conclusions

A new hyperjerk system is constructed and analyzed. This system possesses one quadratic term and single equilibrium point. The dynamical behaviors are investigated by stability analysis, various phase portraits and Lyapunov exponent analysis. The obtained results confirm the chaotic behaviors. As an engineering application, we have derived new synchronization results for the new hyperchaotic hyperjerk system using active backstepping control. In addition, the new hyperjerk have been executed for its simulations using the designed electronical circuit in MultiSIM program. The results of the MultiSIM simulation agree with the numerical simulation.

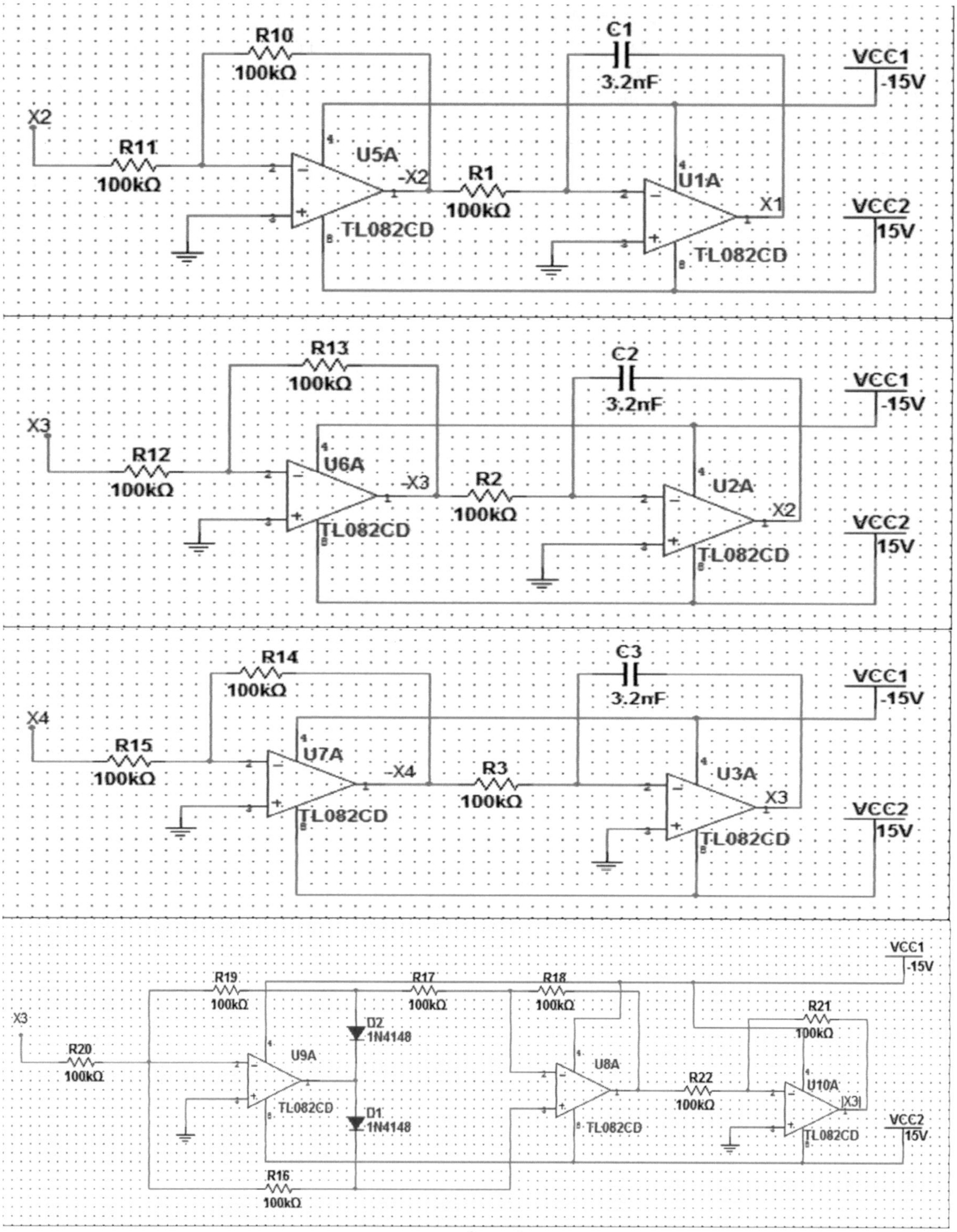
R10
100kΩ
C1
3.2nF
VCC1
-15V
X2
R11
100kΩ
U5A
-X2
R1
100kΩ
U1A
X1
TL082CD
TL082CD
VCC2
15V
R13
100kΩ
C2
3.2nF
VCC1
-15V
X3
R12
100kΩ
U6A
-X3
R2
100kΩ
U2A
X2
TL082CD
TL082CD
VCC2
15V
R14
100kΩ
C3
3.2nF
VCC1
-15V
X4
R15
100kΩ
U7A
-X4
R3
100kΩ
U3A
X3
TL082CD
TL082CD
VCC2
15V
VCC1
-15V
R19
100kΩ
R17
100kΩ
R18
100kΩ
X3
D2
1N4148
R21
100kΩ
R20
100kΩ
U9A
U8A
R22
100kΩ
U10A
|X3|
TL082CD
TL082CD
TL082CD
D1
1N4148
VCC2
15V
R16
100kΩ

continued

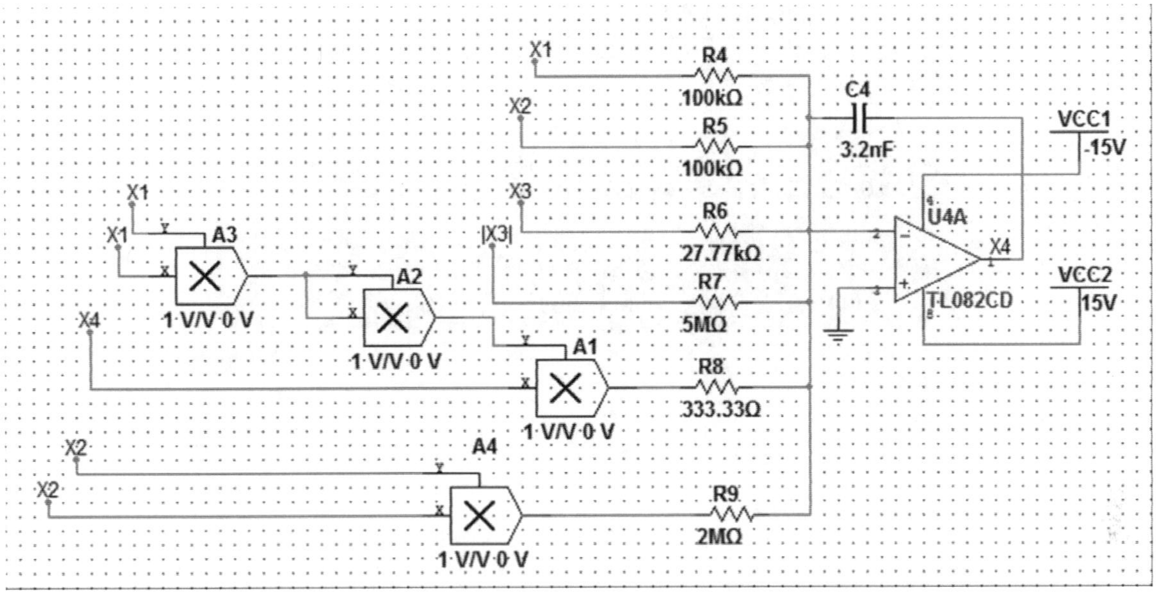

Figure 7. Schematic of the circuit which realizes new hyperjerk system

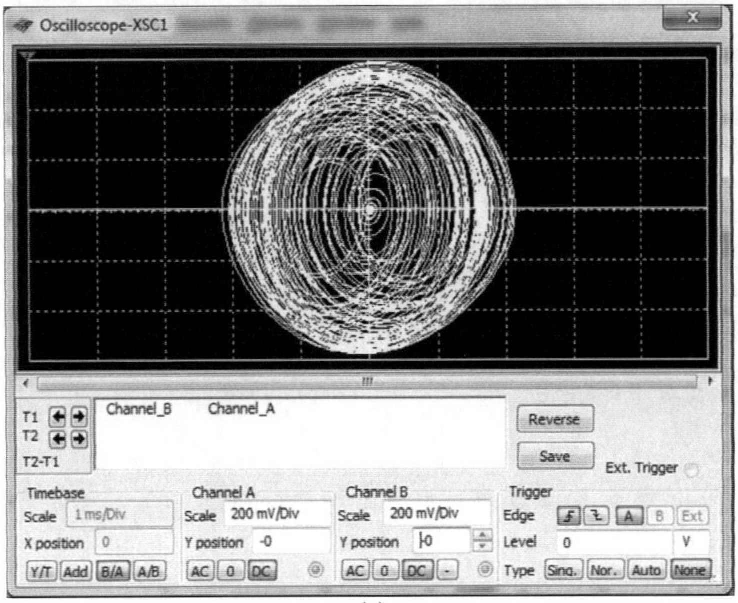

(a)

continued

IOP Conf. Series: Materials Science and Engineering **621** (2019) 012012 doi:10.1088/1757-899X/621/1/012012

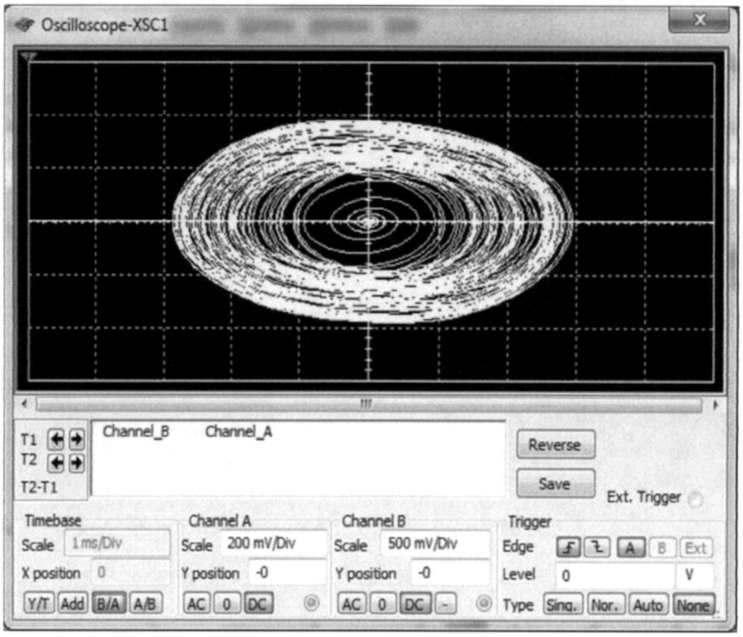

(b)

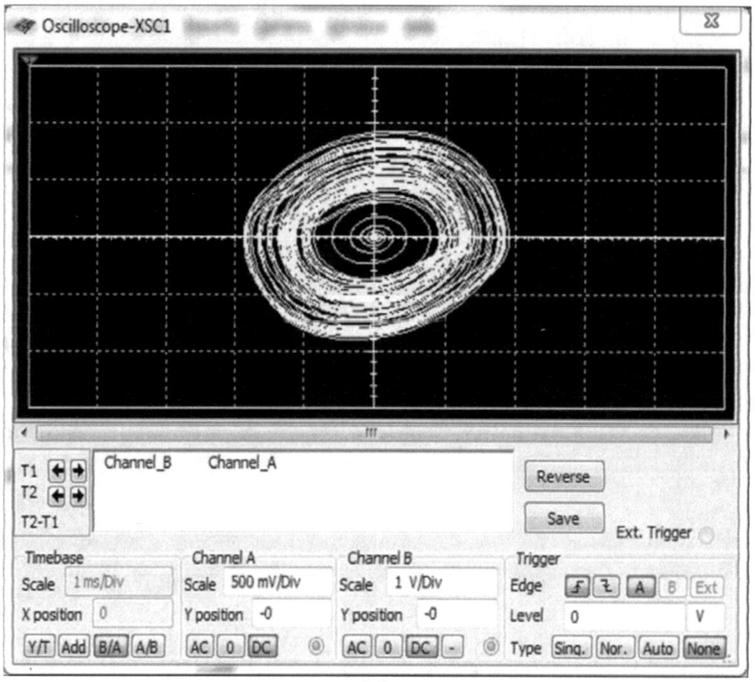

(c)

continued

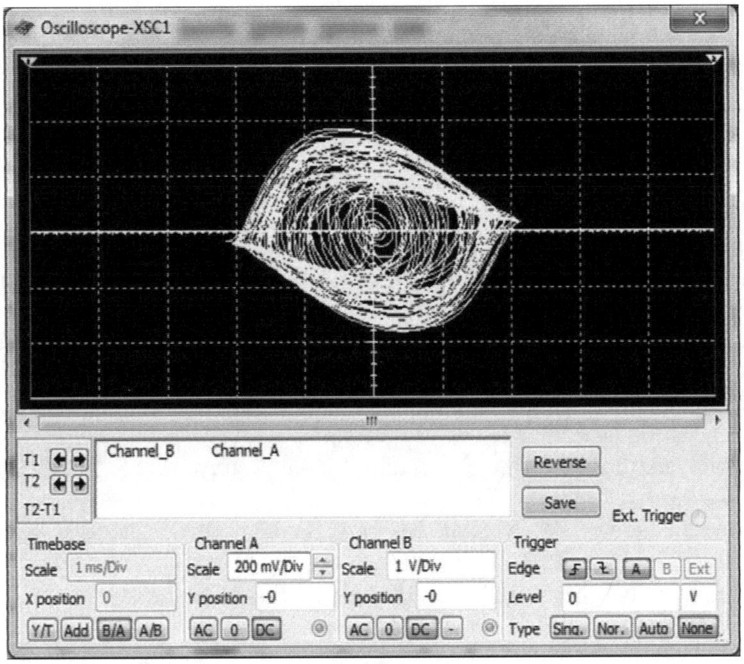

(d)

Figure 9 MultiSIM chaotic attractors in the (a) $x_1 - x_2$ plane, (b) $x_2 - x_3$ plane (c) $x_3 - x_4$ plane and (d) $x_1 - x_4$ plane

References

[1] Azar A T and Vaidyanathan S 2016 Advances in Chaos Theory and Intelligent Control (Berlin: Springer)

[2] Vaidyanathan S and Volos C 2016 Advances and Applications in Chaotic Systems (Berin: Springer)

[3] Vaidyanathan S and Volos C 2016 Advances and Applications in Nonlinear Control Systems (Berlin: Springer)

[4] Pham V T, Vaidyanathan S, Volos C, Kapitaniak T 2018 Nonlinear Dynamical Systems: Self-Excited and Hidden Attractors (Berlin: Springer)

[5] Pham V T, Volos C K and Vaidyanathan S 2015 *Studies in Computational Intelligence* **581** 59-72

[6] Vaidyanathan S, Volos C K and Pham V T 2015 *Studies in Computational Intelligence* **576** 571-590

[7] Vaidyanathan S and Rasappan S 2011 *Communications in Computer and Information Science* **131** 585-593

[8] Vaidyanathan S 2012 *Lecture Notes of the Institute for Computer Sciences, Social-Informatics and Telecommunications Engineering* **85** 124-133 2012

[9] Vaidyanathan S 2015 *Int. J. Modelling, Identification and Control* **23** 380-392

[10] Vaidyanathan S 2015 *Kyunjgpook Mathematical Journal* **55** 563-586

[11] Rasappan S and Vaidyanathan S 2012 *Far East J. Mathematical Sciences* **67** 265-287

[12] Vaidyanathan S, Volos C K, Rajagopal K, Kyprianidis I M and Stouboulos I N 2015 *J. Engineering Science and Technology Review* **8** 74-82

[13] Vaidyanathan S 2015 *Int. J. ChemTech Research* **8** 209-221

[14] Vaidyanathan S 2015 *Int. J. ChemTech Research* **8** 146-158

[15] Vaidyanathan S 2015 *Int. J. ChemTech Research* **8** 159-171
[16] Vaidyanathan S, Sambas A, Mamat M and Sanjaya W S M 2017 *Archives of Control Sciences* **27** 541-54
[17] Mansour S M B, Sundarapandian V and Naceur S M 2016 *International J. Control Theory and Applications* **9** 37-54
[18] Akgul A, Moroz I, Pehlivan I and Vaidyanathan S 2016 *Optik* **127** 5491-99
[19] Zhang Y 2018 *Information Sciences* **450** 361-77
[20] Gunasekaran G and Venkatesan M 2018 *Pertanika J. Science and Technology* **26** 599-614
[21] Pham V T, Jafari S, Volos C, Giakoumis A, Vaidyanathan S and Kapitaniak T 2016 *IEEE Trans. Circuits and Systems II: Express Briefs* **63** 878-82
[22] Volos C, Maaita J O, Vaidyanathan S, Pham V T, Stouboulos I and Kyprianidis I 2017 *IEEE Trans. Circuits and Systems II: Express Briefs* **64** 339-43
[23] Vaidyanathan S, Pham V T, Volos C and Sambas A 2018 *Studies in Systems, Decision and Control* **133** 345-64
[24] Sambas A, Vaidyanathan S, Mamat M and Mada Sanjaya W S 2018 *Studies in Systems, Decision and Control* **133** 365-73
[25] Vaidyanathan S, Sambas A, Mamat M and Sanjaya W S M 2017 *Int. J. Modelling, Identification and Control* 28 153-66
[26] Sambas A, Vaidyanathan S, Mamat M, Sanjaya W S M and Prastio R P 2016 *Int. J. Control Theory and Applications* **9** 141-49
[27] Gottlieb H P W 1996 *American J. Physics* **64** 525
[28] Volos C K, Pham V T, Vaidyanathan S, Kyprianidis I M and Stouboulos I N 2015 *J. Engineering Science and Technology Review* **8** 142-51
[29] Jin J 2018 *Microelectronics Journal* **75** 27-34
[30] Tamba V K, Kuiate G F, Kingni S T and Talla P K 2018 *Studies in Systems, Decisiona nd Control* **133** 203-27
[31] Ghorui S, Sahasrabudhe S N, Murthy P S S, Das A K and Venkataramani N 2000 *IEEE Trans. Plasma Science* **28** 253-60
[32] Vaidyanathan S 2017 *Studies in Computational Intelligence* **709** 435-50
[33] Vaidyanathan S 2016 Studies in Computational Intelligence **636** 137-61
[34] Coullet P, Tresser C and Arneodo A 1979 *Physics Letters A* **72** 268-70
[35] Vaidyanathan S, Volos C, Pham V T and Madhavan K 2015 *Archives of Control Sciences* **25** 135-58
[36] Daltzis P, Vaidyanathan S, Pham V T, Volos C, Nistazakis E and Tombras G 2018 *Circuits, Systems and Signal Processing* **37** 613-35
[37] Chlouveakis K E and Sprott J C 2006 *Chaos, Solitons and Fractals* **28** 739-46
[38] Wang X, Vaidyanathan S, Volos C, Pham V T and Kapitaniak T 2017 *Nonlinear Dynamics* **89** 1673-87
[39] Pham V T, Vaidyanathan S, Volos C, Wang X and Hoang D V 2017 *Studies in Computational Intelligence* **701** 59-80
[40] Vaidyanathan S 2016 *Studies in Computational Intelligence* **635** 185-209
[41] Vaidyanathan S 2012 *Advances in Intelligent Systems and Computing* **176** 329-37
[42] Vaidyanathan S and Pakiriswamy S 2012 *Lecture Notes of the Institute for Computer Sciences, Social-Informatics and Telecommunications Engineering* **84** 111-18
[43] Vaidyanathan S, Azar A T and Boulkroune 2018 *Int. J. Automation and Control* **12** 5-26
[44] Oumate A A, Vaidyanathan S, Zourmba K, Gambo B and Mohamadou A 2018 Studies in Systems, Decision and Control 133 251-70
[45] Vaidyanathan S 2017 *Studies in Computational Intelligence* **701** 393-423
[46] Daltzis P A, Volos C K, Nistazakis H E, Tsigopoulos A D and Tombras G S 2018 *Computation* **6** 14.
[47] Vaidyanathan S 2016 *Archives of Control Sciences* **26** 311-38

[48] Vaidyanathan S 2017 *Archives of Control Sciences* **27** 409-39
[49] Wolf A, Swift J B, Swinney H L and Vastano J A 1985 *Physica D* **16** 285-317
[50] Khalil H K 2001 Nonlinear Systems (New York: Pearson)
[51] Sambas A, Mujiarto, Mamat M, Sanjaya W S M 2017 *Far East Journal of Mathematical Sciences* **102** 1165-1177
[52] Mamat M, Vaidyanathan S, Sambas A, Mujiarto, Sanjaya W S M, Subiyanto 2018 *IOP Conference Series: Materials Science and Engineering* **332** 012033.
[53] Lien C H, Vaidyanathan S, Sambas A, Sukono, Mamat M, Sanjaya W S M, Subiyanto *IOP Conference Series: Materials Science and Engineering* **332** 012010

IOP Conf. Series: Materials Science and Engineering **621** (2019) 012013 doi:10.1088/1757-899X/621/1/012013

A new 3-D chaotic system with four quadratic nonlinear terms, its global chaos control via passive control method and circuit design

***C H Lien[1], S Vaidyanathan[2], A Sambas[3], Sampath S[4], Sukono[5], M Mamat[6]**

[1]Department of Marine Engineering, National Kaohsiung University of Science and Technology, Kaohsiung, TAIWAN

[2]Research and Development Centre, Vel Tech University, Chennai, Tamil Nadu, INDIA

[3]Department of Mechanical Engineering, Universitas Muhammadiyah Tasikmalaya, INDONESIA

[4]Department of Electronics and Communication Engineering, Vel Tech University, Chennai, Tamil Nadu, INDIA

[5]Department of Mathematics, Faculty of Mathematics and Natural Sciences,

Universitas Padjadjaran, INDONESIA

[6]Faculty of Informatics and Computing, Universiti Sultan Zainal Abidin, KualaTerengganu, MALAYSIA

Corresponding author's e-mail address: chlien@mail.nkmu.edu.tw

Abstract. This paper reports the finding of a new three-dimensional chaotic system with four quadratic nonlinear terms. The paper starts with a detailed dynamic analysis of the properties of the system such as phase plots, Lyapunov exponents, Kaplan-Yorke dimension and equilibrium points. Our new chaotic system is obtained by modifying the dynamics of the Zhu chaotic system (2010), and it has complex chaotic properties. As an engineering application, passive control method is applied for the global chaos control of the new chaotic system. Finally, an electronic circuit implementation of the new chaotic system is designed and implemented in MultiSIM. A good qualitative agreement has been shown between the MATLAB simulations of the new chaotic system and the MultiSIM results.

1. Introduction

Chaotic dynamical systems have applications in several areas of science and engineering [1-4]. Chaotic systems are highly complex and sensitive to even small changes in the initial conditions of the systems. Chaotic systems find applications in areas such as oscillators [5-10], ecology [11-12], biology [13-15], fuzzy logic [16-17], artificial neural networks [18-20], weather systems [21-22], chemical reactors [23-25], robotics [26-27], communication systems [28-30], circuits [31-36], finance systems [37-38], jerk systems [39-42], hyperjerk systems [43-45], etc.

In this research work, we have derived a new chaotic system by modifying the dynamics of the Zhu chaotic system ([46], 2010). Specifically, our new chaotic system has four quadratic nonlinear terms and it exhibits more chaotic properties than the Zhu chaotic system. By deriving Lyapunov exponents of the new chaotic system, we have done a detailed qualitative study of the new chaotic system such as its phase plots, maximum Lyapunov exponent, Kaplan-Yorke dimension, equilibrium points, etc.

As an engineering application, we have used passive control [47-48] for the global chaos control of the new chaotic system. We have established our main result using passivity-based control via Lyapunov stability theory.

The organization structure of this paper is as follows. Section 2 announces the new chaotic system with four quadratic nonlinear terms and details its qualitative properties. Section 3 details the global chaos control of the new chaotic system via passive control. Section 4 describes an electronic circuit design of the new chaotic jerk system using MultiSIM. Section 5 draws the main conclusions.

2. A new chaotic system with four quadratic nonlinearities

In 2010, Zhu, Lio and Guo [46] proposed a 3-D chaotic system given by the dynamics

$$\begin{cases} \dot{x}_1 = -x_1 - a x_2 + x_2 x_3 \\ \dot{x}_2 = b x_2 - x_1 x_3 \\ \dot{x}_3 = -c x_3 + x_1 x_2 \end{cases} \tag{1}$$

In Eq. (1), x_1, x_2, x_3 are the states and a, b, c are positive parameters. In [46], it was shown that the Zhu system (1) is chaotic for the choice of parameters $(a,b,c) = (1.5, 2.5, 4.9)$.

Using Wolf's algorithm [49], the Lyapunov exponents of the Zhu system (1) are calculated for $(a,b,c) = (1.5, 2.5, 4.9)$ and $X(0) = (0.2, 0.2, 0.2)$ for $T = 1E4$ seconds as

$$L_1 = 0.6616, \quad L_2 = 0, \quad L_3 = -4.0662 \tag{2}$$

The Zhu system (1) is chaotic as it possesses a positive Lyapunov exponent in (2).

The Kaplan-Yorke fractal dimension of the Zhu system (1) is calculated as

$$D_{KY} = 2 + \frac{L_1 + L_2}{|L_3|} = 2.1627 \tag{3}$$

The value $D_{KY} = 2.1627$ describes the chaotic nature of the Zhu system (1).

In this paper, we propose a new chaotic system by modifying the dynamics of the Zhu system (1) and obtaining the following system:

$$\begin{cases} \dot{x}_1 = -a(x_1 + x_2) + x_2 x_3 \\ \dot{x}_2 = b x_2 - x_1 x_3 - d x_1^2 \\ \dot{x}_3 = -c x_3 + x_1 x_2 \end{cases} \tag{4}$$

In Eq. (4), x_1, x_2, x_3 are the states and a, b, c are positive parameters.

We show that the system (4) is chaotic for the parameter values

$$a = 1.2, \quad b = 3, \quad c = 5, \quad d = 0.1 \tag{5}$$

Using Wolf's algorithm [49], the Lyapunov exponents of the new chaotic system (4) are obtained for the parameter values as in (7) and $X(0) = (0.1, 0.1, 0.1)$ for $T = 1E4$ seconds as

$$L_1 = 0.7248, \quad L_2 = 0, \quad L_3 = -3.9289 \tag{6}$$

Clearly, the new system (4) is chaotic since it has a positive Lyapunov exponent L_1.

By adding all the Lyapunov exponents in (6), we get their sum as

IOP Conf. Series: Materials Science and Engineering **621** (2019) 012013 doi:10.1088/1757-899X/621/1/012013

$$L_1 + L_2 + L_3 = -3.2041 < 0 \tag{7}$$

Thus, the new chaotic system (4) is dissipative and it has a strange chaotic attractor.

The Kaplan-Yorke fractal dimension of the new chaotic system (4) is calculated as

$$D_{KY} = 2 + \frac{L_1 + L_2}{|L_3|} = 2.1845 \tag{8}$$

The maximal Lyapunov exponent (MLE) of the new chaotic system (4) is found as $L_1 = 0.7248$, which is greater than the maximal Lyapunov exponent (MLE) of the Zhu chaotic system (1) found as $L_1 = 0.6616$. Similarly, the Kaplan-Yorke dimension of the new chaotic system (4) is seen as $D_{KY} = 2.1845$, which is greater than the Kaplan-Yorke dimension of the Zhu chaotic system (1) given by $D_{KY} = 2.1627$.

The equilibrium points of the new chaotic system (4) are calculated by solving the equations

$$-a(x_1 + x_2) + x_2 x_3 = 0 \tag{9a}$$

$$bx_2 - x_1 x_3 - dx_1^2 = 0 \tag{9b}$$

$$-cx_3 + x_1 x_2 = 0 \tag{9c}$$

Solving the system (9) for the chaotic case $(a,b,c,d) = (1.2, 3, 5, 0.1)$, we get five equilibrium points of the new chaotic system (4), listed as follows:

$$E_0 = \begin{bmatrix} 0 \\ 0 \\ 0 \end{bmatrix}, \; E_1 = \begin{bmatrix} 3.6173 \\ 3.4154 \\ 2.4709 \end{bmatrix}, \; E_2 = \begin{bmatrix} 4.5229 \\ -1.8744 \\ -1.6956 \end{bmatrix}, \; E_3 = \begin{bmatrix} -3.4097 \\ 1.7229 \\ -1.1749 \end{bmatrix}, \; E_4 = \begin{bmatrix} -4.2087 \\ -3.2639 \\ 2.7474 \end{bmatrix} \tag{10}$$

It is easy to verify that E_0 is a saddle point and other equilibrium points are saddle-foci.

Thus, all equilibrium points of the new chaotic system are unstable.

Figures 1-3 show the 2-D phase portraits of the new chaotic system (4), while Figure 4 shows the 3-D phase portrait. Figure 5 shows the Lyapunov exponents of the new chaotic system (4).

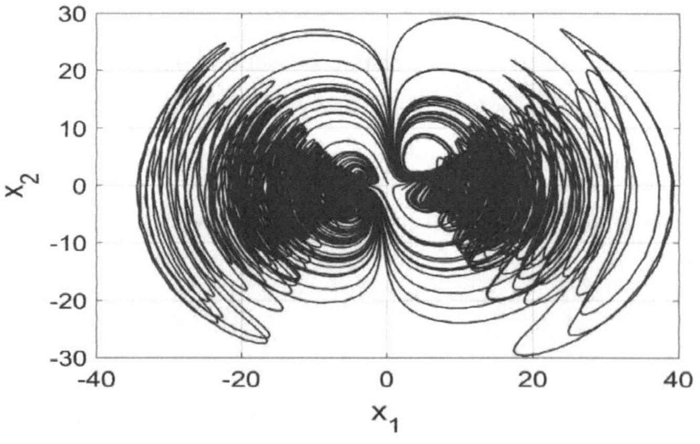

Figure 1. Two-dimensional phase plot of the new chaotic system (4) in the $(x_1, x_2)-$ plane for $(a,b,c,d) = (1.2, 3, 5, 0.1)$ and $X(0) = (0.2, 0.2, 0.2)$

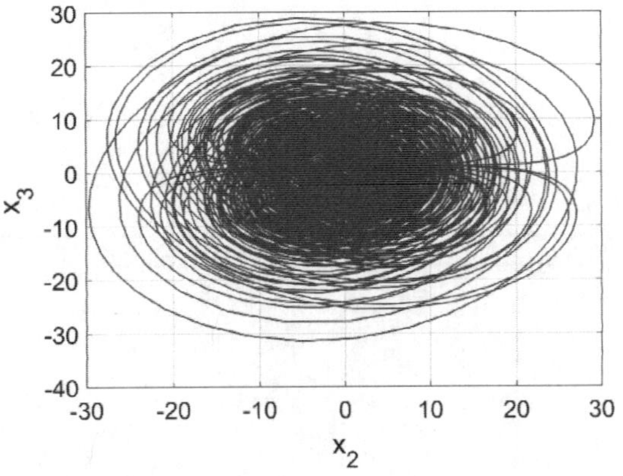

Figure 2. Two-dimensional phase plot of the new chaotic system (4) in the $(x_2, x_3)-$ plane for $(a,b,c,d) = (1.2, 3, 5, 0.1)$ and $X(0) = (0.2, 0.2, 0.2)$

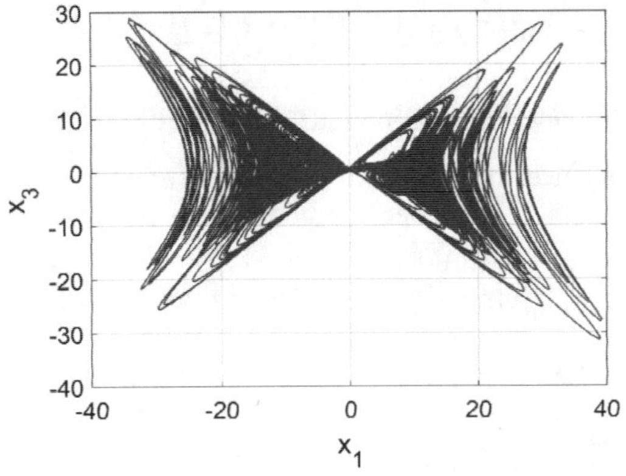

Figure 3. Two-dimensional phase plot of the new chaotic system (4) in the $(x_1, x_3)-$ plane for $(a,b,c,d) = (1.2, 3, 5, 0.1)$ and $X(0) = (0.2, 0.2, 0.2)$

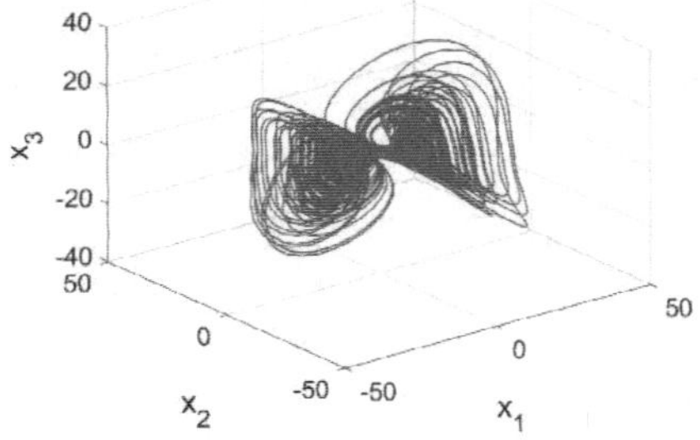

Figure 4. Strange attractor of the new chaotic system (4) in R^3 for $(a,b,c,d)=(1.2,3,5,0.1)$ and $X(0)=(0.2,0.2,0.2)$

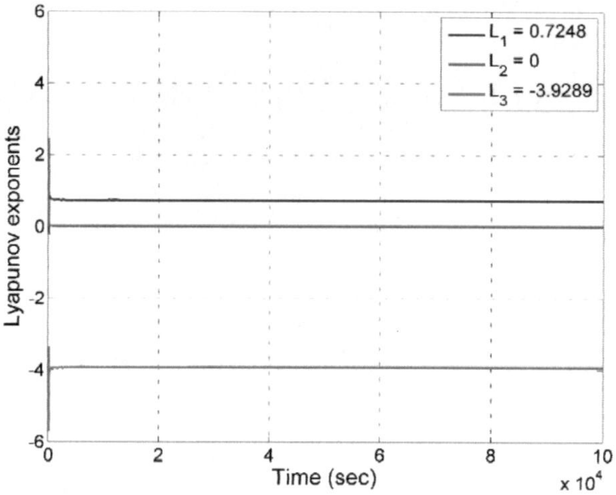

Figure 5. Lyapunov exponents of the new chaotic system (4) for $(a,b,c,d)=(1.2,3,5,0.1)$ and $X(0)=(0.2,0.2,0.2)$

3. Global chaos control of the new chaotic system via passive control method

In this section, we exhibit an engineering application of the new chaotic system (4), namely global chaos control of the new chaotic system (4) via passive control method. The main result in this section is established via Lyapunov stability theory [50].

Thus, we consider the controlled chaotic system given by

$$\begin{cases} \dot{x}_1 = -a(x_1 + x_2) + x_2 x_3 \\ \dot{x}_2 = b x_2 - x_1 x_3 - d x_1^2 + u \\ \dot{x}_3 = -c x_3 + x_1 x_2 \end{cases} \tag{11}$$

In (11), x_1, x_2, x_3 are states and u is a passive control to be determined.

We suppose that x_2 is the output of the new chaotic system (11), *i.e.* $y = x_2$.

We suppose also that $z_1 = x_1$ and $z_2 = x_3$.

Then the control system (11) can be expressed in normal form as follows:

$$\begin{cases} \dot{z}_1 = -a z_1 + (z_2 - a) y \\ \dot{z}_2 = -c z_2 + z_1 y \\ \dot{y} = b y - z_1 z_2 - d z_1^2 + u \end{cases} \tag{12}$$

We express the system (12) in the standard form of passive control theory [48] as

$$\begin{cases} \dot{z} = f_0(z) + p(z, y)y \\ \dot{y} = \beta(z, y) + \alpha(z, y)u \end{cases} \tag{13}$$

where

$$f_0(z) = \begin{bmatrix} -az_1 \\ -cz_2 \end{bmatrix}, \quad p(z, y) = \begin{bmatrix} z_2 - a \\ z_1 \end{bmatrix}, \quad \alpha(z, y) = 1, \quad \beta(z, y) = by - z_1 z_2 - dz_1^2 \tag{14}$$

Now, we consider the quadratic positive definite function defined by

$$W(z) = \frac{1}{2}\left(z_1^2 + z_2^2\right) \tag{15}$$

We claim that $W(z)$ is a Lyapunov function for the dynamics

$$\dot{z} = f_0(z) = \begin{bmatrix} -az_1 \\ -cz_2 \end{bmatrix} \tag{16}$$

Taking the time-derivative of W along the dynamics (16), we get

$$\dot{W}(z) = \frac{\partial W}{\partial z} f_0(z) = -az_1^2 - cz_2^2 \tag{17}$$

this is a negative definite function on R^2.

As in [48], we choose the storage function as

$$V(z, y) = W(z) + \frac{1}{2} y^2 \tag{18}$$

We note that V is a quadratic and positive definite function on R^3.

Taking the time-derivative of V along the trajectories of the system (13), we get

$$\dot{V}(z) = \frac{\partial W}{\partial z} f_0(z) + \frac{\partial W}{\partial z} p(z, y)y + \left[\beta(z, y) + \alpha(z, y)u\right]y \tag{19}$$

A simple calculation gives

$$\dot{V} = -az_1^2 - cz_2^2 + y\left[z_1(z_2 - a) - dz_1^2 + by + u\right] \tag{20}$$

We take the feedback control u as

$$u = -z_1(z_2 - a) + dz_1^2 - (b + k)y \tag{21}$$

where $k > 0$ is a gain constant.

Substituting (21) into (20), we get

$$\dot{V} = -az_1^2 - cz_2^2 - ky^2 \tag{22}$$

this is a quadratic and negative definite function on R^3.

Substituting $y = x_2, z_1 = x_1$ and $z_2 = x_3$, we can also express the passive control (21) as

$$u = -x_1(x_3 - a) + dx_1^2 - (b+k)x_2 \tag{23}$$

By Lyapnov stability theory [50], we have established the following main result.

Theorem 1. The new chaotic system (11) is globally and exponentially stabilized for all initial conditions $x(0) \in R^3$ by the passive control u defined by Eq. (23), where $k > 0$ is a gain constant. ■

For numerical simulations, we take the gain constant as $k = 20$. We take the parameter values as in the chaotic case, *viz.* $(a,b,c,d) = (1.2, 3, 5, 0.1)$.

As the initial conditions, we take $(x_1(0), x_2(0), x_3(0)) = (5.4, 12.8, 3.9)$.

Figure 6 shows the time-history of the trajectories of the controlled system (11), when the passive control u given by Eq. (23) is activated.

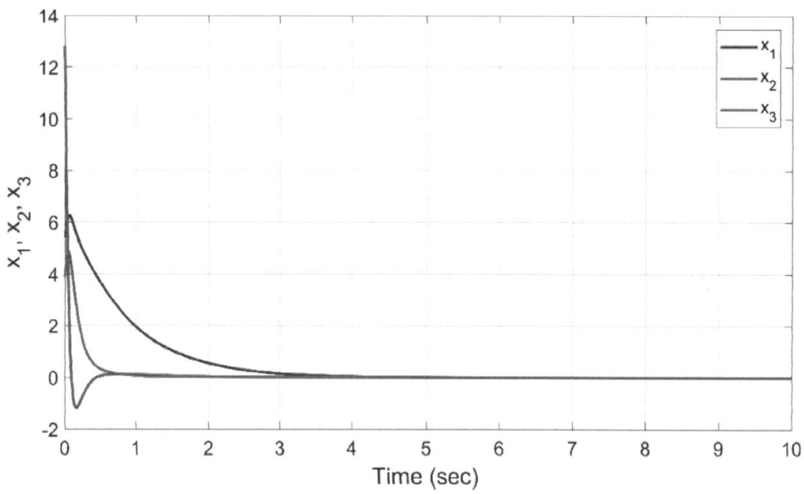

Figure 6. Time-history of the controlled chaotic system (11) for the initial state $X(0) = (5.4, 12.8, 3.9)$, parameter values $(a,b,c,d) = (1.2, 3, 5, 0.1)$ and $k = 20$

4. Circuit realization of the new chaotic system

In this section, we will design an electronic circuit for new three-dimensional chaotic system with four quadratic nonlinear terms. The circuit in Figure 7 is designed using an approach based operational amplifiers where the state variables x_1, x_2 and x_3 of new chaotic system (4) are associated with the voltages across the capacitors C_1, C_2 and C_3, respectively. Four multipliers (AD633JN), resistors, capacitors and five amplifiers (TL082CD) are used to design the system dynamics in the circuit.

In accordance with the circuit schematic, the new chaotic system (4) can be rewritten as follows:

$$\begin{cases} \dot{x}_1 = -\dfrac{1}{C_1R_1}x_1 - \dfrac{1}{C_1R_2}x_2 + \dfrac{1}{10C_1R_3}x_2x_3 \\ \dot{x}_2 = \dfrac{1}{C_2R_4}x_2 - \dfrac{1}{10C_2R_5}x_1x_3 - \dfrac{1}{10C_2R_6}x_1^2 \\ \dot{x}_3 = -\dfrac{1}{C_3R_7}x_3 + \dfrac{1}{10C_3R_8}x_1x_2 \end{cases} \tag{24}$$

MultiSIM results are shown in Figure 8 by choosing convenient values for resistors and capacitors: $R_1 = R_2 = 333.33$ kΩ, $R_3 = R_5 = R_8 = 40$ kΩ, $R_4 = 133.3$ kΩ, $R_6 = 400$ kΩ, $R_7 = 80$ kΩ, $R_9 = R_{10} = R_{11} = R_{12}$ =100 kΩ, $C_1 = C_2 = C_3 = C_4 = 3.2$ nF. Figure 8 presents MultiSIM results of the circuit, which show its chaotic behavior.

5. Conclusions

a new three-dimensional chaotic system with four quadratic nonlinear terms has been proposed in this paper. Numerical simulations have been used to illustrate that this system has a chaotic behavior. The dynamical properties of this new chaotic system were analyzed. As an engineering application, we have derived new results for the global chaos control of the new chaotic system by passivity based control. The designed circuit has been implemented and examined using the MultiSIM software to verify the simulation results.

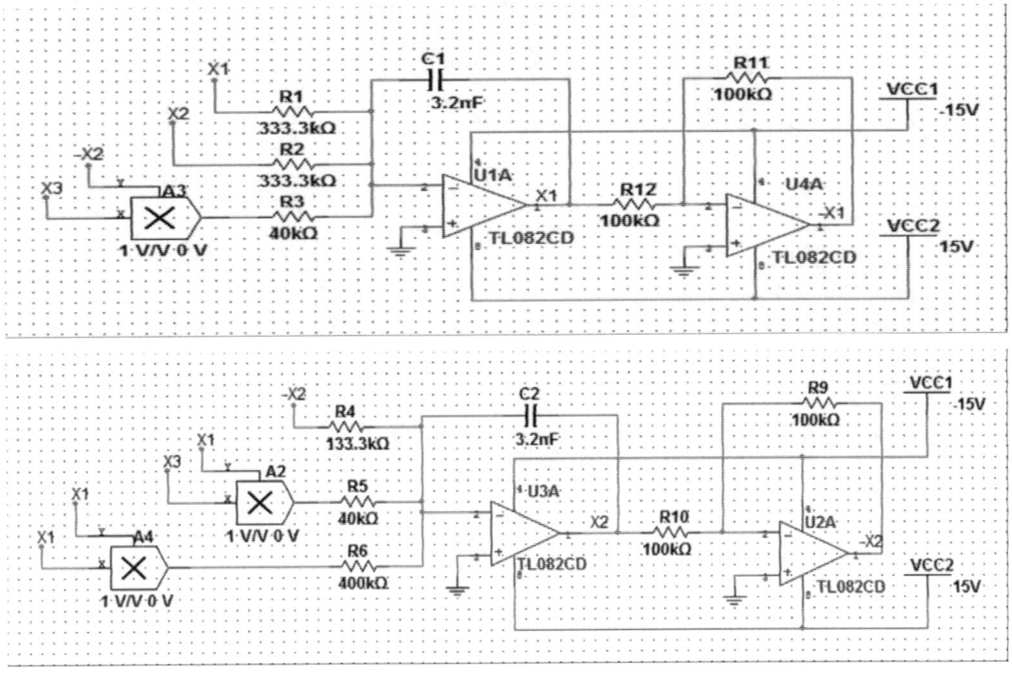

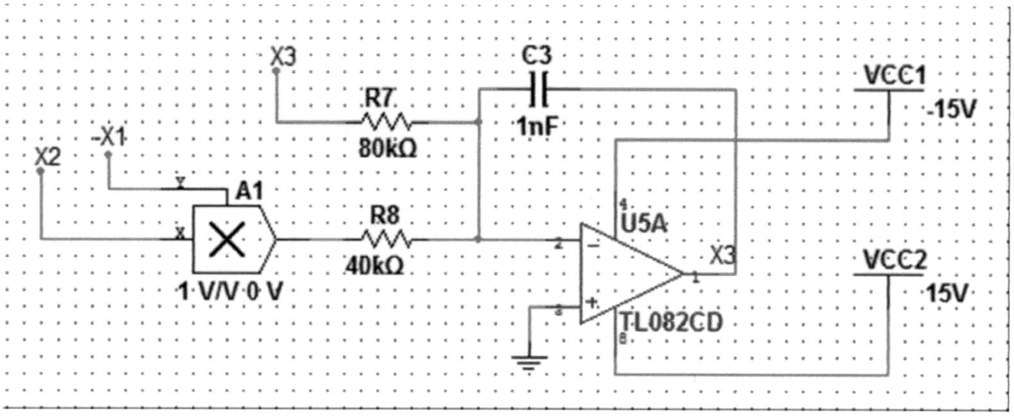

Figure 7. The electronic circuit schematic of the new chaotic system (4)

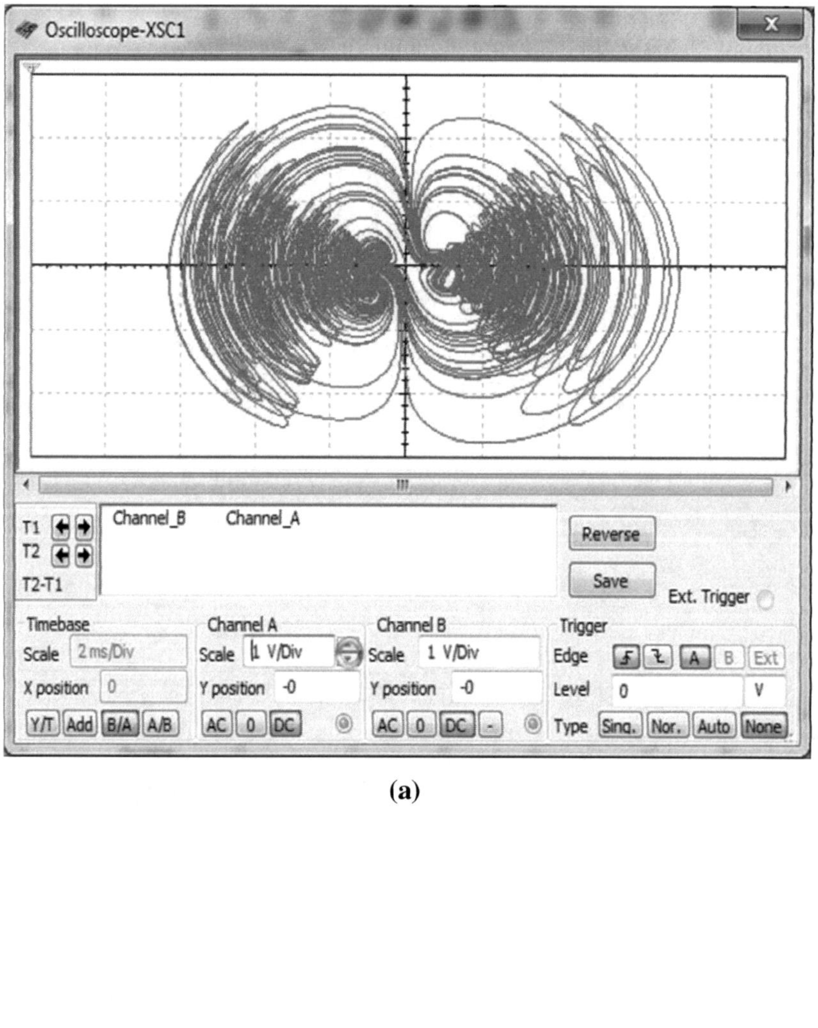

(a)

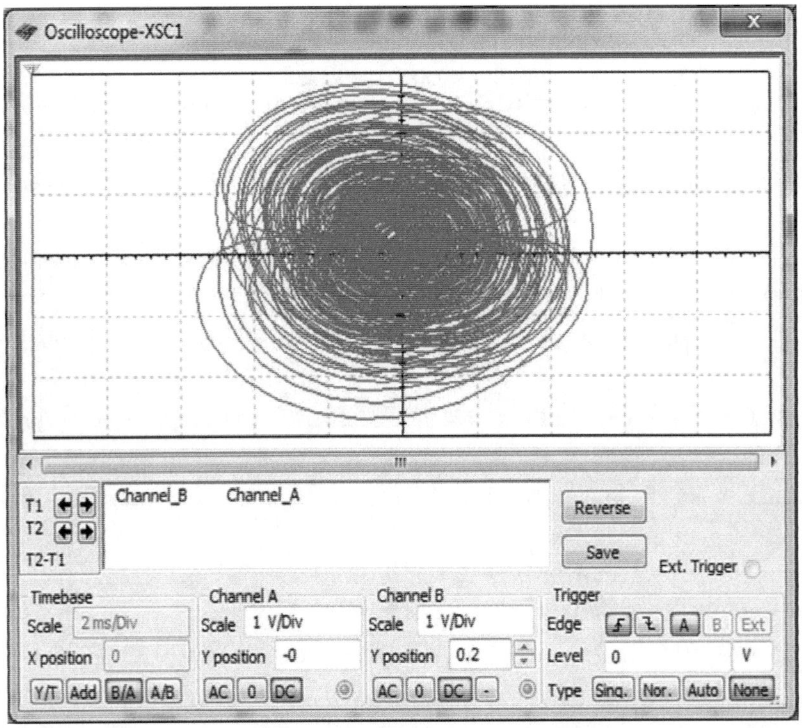

(b)

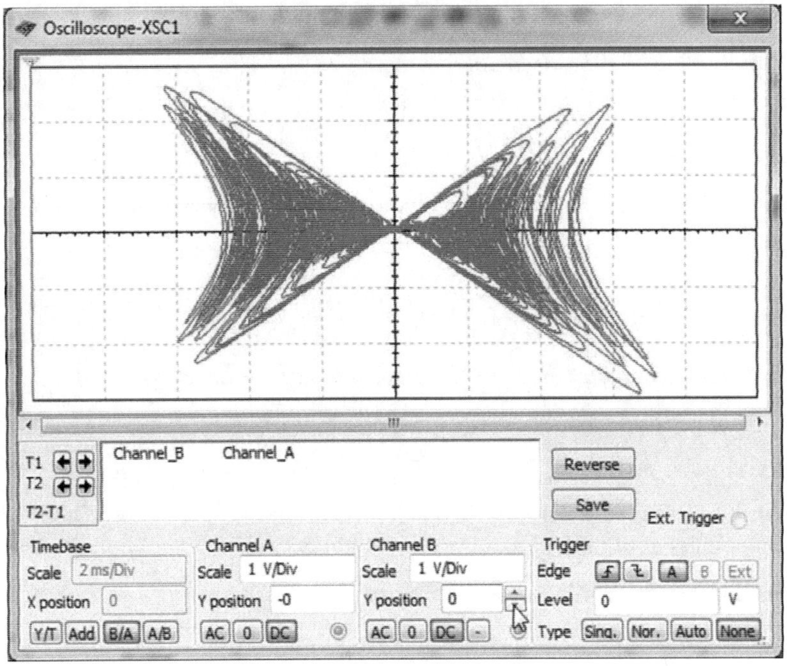

(c)

Figure 8. Phase portraits of the MultiSIM simulation in the (a) $x_1 - x_2$ plane, (b) $x_2 - x_3$ plane and (c) $x_1 - x_3$ plane

References

[1] Azar A T and Vaidyanathan S 2016 Advances in Chaos Theory and Intelligent Control (Berlin: Springer)

[2] Vaidyanathan S and Volos C 2016 Advances and Applications in Chaotic Systems (Berin: Springer)

[3] Vaidyanathan S and Volos C 2016 Advances and Applications in Nonlinear Control Systems (Berlin: Springer)

[4] Pham V T, Vaidyanathan S, Volos C, Kapitaniak T 2018 Nonlinear Dynamical Systems: Self-Excited and Hidden Attractors (Berlin: Springer)

[5] Pham V T, Volos C K and Vaidyanathan S 2015 *Studies in Computational Intelligence* 581 59-72

[6] Vaidyanathan S, Volos C K and Pham V T 2015 *Studies in Computational Intelligence* 576 571-590

[7] Vaidyanathan S and Rasappan S 2011 *Communications in Computer and Information Science* 131 585-593

[8] Vaidyanathan S 2012 *Lecture Notes of the Institute for Computer Sciences, Social-Informatics and Telecommunications Engineering* 85 124-133 2012

[9] Vaidyanathan S 2015 *Int. J. Modelling, Identification and Control* 23 380-92

[10] Vaidyanathan S 2015 *Kyunjgpook Mathematical Journal* 55 563-586

[11] Matouk A E and Elsadany A A 2016 *Nonlinear Dynamics* 85 1597-612

[12] Chattopadhyay J, Pal N, Samanta S, Venturino E and Khan Q J A 2015 *BioSystems* 138 18-24

[13] Saha S and Mukherjee V 2018 *Soft Computing* 22 3797-816

[14] Nikolaev E V, Rahi S J and Sontag E D 2018 *Biophysical Journal* 114 1232-40

[15] Scharf Y 2017 *Chaos Solitons Fractals* 95 42-47

[16] Vaidyanathan S and Azar A T 2016 *Int. J. Intelligent Engineering Informatics* 4 135-50

[17] Shirkavand M and Pourgholi M 2017 *Chaos Solitons Fractals* 113 135-47

[18] Njitacke Z T and Kengne J 2018 *AEU-International Journal of Electronics and Communications* 93 245-52

[19] Tooranjipour P and Vatankhah R 2018 *Applied Soft Computing Journal* 70 622-32

[20] Xu C 2018 *Mathematics and Computers in Simulation* 149 69-90

[21] Rasappan S and Vaidyanathan S 2012 *Far East J. Mathematical Sciences* 67 265-287

[22] Vaidyanathan S, Volos C K, Rajagopal K, Kyprianidis I M and Stouboulos I N 2015 *J. Engineering Science and Technology Review* 8 74-82

[23] Vaidyanathan S 2015 *Int. J. ChemTech Research* 8 209-221

[24] Vaidyanathan S 2015 *Int. J. ChemTech Research* 8 146-158

[25] Vaidyanathan S 2015 *Int. J. ChemTech Research* 8 159-171

[26] Vaidyanathan S, Sambas A, Mamat M and Sanjaya W S M 2017 *Archives of Control Sciences* 27 541-54

[27] Mansour S M B, Sundarapandian V and Naceur S M 2016 *International J. Control Theory and Applications* 9 37-54

IOP Conf. Series: Materials Science and Engineering **621** (2019) 012013 doi:10.1088/1757-899X/621/1/012013

[28] Akgul A, Moroz I, Pehlivan I and Vaidyanathan S 2016 *Optik* 127 5491-99

[29] Zhang Y 2018 *Information Sciences* 450 361-77

[30] Gunasekaran G and Venkatesan M 2018 *Pertanika J. Science and Technology* 26 599-614

[31] Pham V T, Jafari S, Volos C, Giakoumis A, Vaidyanathan S and Kapitaniak T 2016 *IEEE Trans. Circuits and Systems II: Express Briefs* 63 878-82

[32] Volos C, Maaita J O, Vaidyanathan S, Pham V T, Stouboulos I and Kyprianidis I 2017 *IEEE Trans. Circuits and Systems II: Express Briefs* 64 339-43

[33] Vaidyanathan S, Pham V T, Volos C and Sambas A 2018 *Studies in Systems, Decision and Control* 133 345-64

[34] Sambas A, Vaidyanathan S, Mamat M and Mada Sanjaya W S 2018 *Studies in Systems, Decision and Control* 133 365-73

[35] Vaidyanathan S, Sambas A, Mamat M and Sanjaya W S M 2017 *Int. J. Modelling, Identification and Control* 28 153-66

[36] Sambas A, Vaidyanathan S, Mamat M, Sanjaya W S M and Prastio R P 2016 *Int. J. Control Theory and Applications* 9 141-49

[37] Idowu B A, Vaidyanathan S, Sambas A, Olusola O I and Onma O S 2018 *Studies in Systems, Decision and Control* 133 271-95

[38] Tacha O I, Volos C K, Kyprianidis I M, Stouboulos I N, Vaidyanathan S and Pham V T 2016 *Applied Math. Computation* 276 200-17

[39] Vaidyanathan S 2017 *Archives of Control Sciences* 27 409-39

[40] Vaidyanathan S, Sambas A, Mamat M and Sanjaya W S M 2017 *Int. J. Modelling, Identification and Control* 28 153-66

[41] Vaidyanathan S 2016 *Int. J. Control Theory and Applications* 9 199-219

[42] Vaidyanathan S 2016 *Studies in Computational Intelligence* 635 59-79

[43] Wang X, Vaidyanathan S, Volos C, Pham V T and Kapitaniak T 2017 *Nonlinear Dynamics* 89 1673-87

[44] Leutcho G D, Kengne J and Kengne L K 2018 *Chaos Solitons Fractals* 107 67-87

[45] Daltzis P, Vaidyanahtan S, Pham V T, Volos C, Nistazakis E and Tombras G 2018 *Circuits Systems and Signal Processing* 37 613-35

[46] Zhu C, Liu Y and Guo Y 2010 *Intelligent Information Management* 2 104-9

[47] Kose E 2017 *Electrical Engineering* 99 763-73

[48] Byrnes C I, Isidori I and Willems J C 1991 *IEEE Transactions on Automatic Control* 36 1228-40

[49] Wolf A, Swift J B, Swinney H L and Vastano J A 1985 *Physica D* 16 285-317

[50] Khalil H K 2001 Nonlinear Systems (New York: Pearson)

IOP Conf. Series: Materials Science and Engineering **621** (2019) 012014 doi:10.1088/1757-899X/621/1/012014

A New 4-D Hyperchaotic System with Four-Scroll Hidden Attractor, Its Properties and Bifurcation Analysis

L G Dolvis[1], S Vaidyanathan[2], K Jacques[3] A Sambas[4], Sukono[5], M Mamat[6],

[1]Laboratory of Condensed Matter, Electronics and Signal Processing, Faculty of Science, Department of Physics, University of Dschang, Dschang, CAMEROON
[2]Research and Development Centre, Vel Tech University, Chennai, Tamil Nadu, INDIA
[3]Department of Electrical Engineering, IUT FV, Bandjoun, University of Dschang, Dschang, CAMEROON
[4]Department of Mechanical Engineering, Universitas Muhammadiyah Tasikmalaya, INDONESIA
[5]Department of Mathematics, Faculty of Mathematics and Natural Sciences Universitas Padjadjaran, INDONESIA
[6]Faculty of Informatics and Computing, Universiti Sultan Zainal Abidin, KualaTerengganu, MALAYSIA

Corresponding author's e-mail address: leutchoeinstein@yahoo.com

Abstract. This paper announces a new four-dimensional hyperchaotic system with a four-scroll attractor and discusses its dynamic properties such as Lyapunov exponents, phase portraits, Kaplan-Yorke dimension and equilibrium points. Our calculations show that the new hyperchaotic system has no equilibrium point and hence it exhibits hidden attractor. Our new hyperchaotic system has three nonlinearities in total. A detailed bifurcation analysis has been presented for the new hyperchaotic system with four-scroll hidden attractor. Specifically, we discussed bifurcation analysis such as route to four-scroll hyperchaos, coexisting bifurcation, multistability, two parameter Lyapunov exponents and antimonotonicity.

1. Introduction

Chaotic and hyperchaotic dynamical systems have applications in several areas of science and engineering [1-4]. Some popular applications of chaotic systems can be cited as oscillators [5-10], weather systems [11-12], chemical reactors [13-15], robotics [16-17], communication systems [18-20], circuits [21-26], etc. Hyperchaotic systems are usually classified as chaotic systems having more than one positive Lyapunov exponent. The first hyperchaotic system was reported by Rössler in 1976 [27]. Hyperchaotic systems have many applications such as secure communication devices [28-30], memristors [31-33], cryptosystems [34-35], encryption [36-40], etc.
In this paper, we report the finding of a new 4-D hyperchaotic system with a four-scroll attractor. Our new hyperchaotic system has three quadratic nonlinearities. It is interesting to note that our new hyperchaotic system has no equilibrium point. Hence it exhibits hidden attractor [41-43]. In the literature, there is great interest in finding hyperchaotic systems with hidden attractor [44-45].

Section 2 announces the new hyperchaotic system with four-scroll attractor and no equilibrium point. Section 3 details bifurcation analysis of the new hyperchaotic system. Specifically, we give a detailed bifurcation analysis such as route to four-scroll hyperchaos, coexisting bifurcation, multistability, two parameter Lyapunov exponents and antimonotonicity. Section 4 draws the main conclusions of this paper.

IOP Conf. Series: Materials Science and Engineering **621** (2019) 012014 doi:10.1088/1757-899X/621/1/012014

2. A new 4-D hyperchaotic system with four-scroll hidden attractor

In this work, we announce a new 4-D system with the following dynamics

$$\begin{cases} \dot{x} = a(x-y) - yz + w \\ \dot{y} = xz - by + w \\ \dot{z} = xy - cz \\ \dot{w} = -x \end{cases} \tag{1}$$

In Eq. (1), x, y, z, w are the states and a, b, c are positive parameters. In this work, we show that the system (1) is hyperchaotic for the parameter values

$$a = 7, \quad b = 30, \quad c = 10 \tag{2}$$

Using Wolf's algorithm [46], the Lyapunov exponents of the new hyperchaotic system (1) are calculated for $(a,b,c) = (7,30,10)$ and $X(0) = (0.1, 0.1, 0.1, 0.1)$ for $T = 1E4$ seconds as

$$L_1 = 2.2703, \quad L_2 = 0.0794, \quad L_3 = 0, \quad L_4 = -30.5747 \tag{3}$$

Since the Lyapunov exponents L_1 and L_2 are positive in (6), we conclude that the new system (1) is hyperchaotic. Also, we calculate the Kaplan-Yorke dimension of the new hyperchaotic system (1) as

$$D_{KY} = 3 + \frac{L_1 + L_2 + L_3}{|L_4|} = 3.0769 \tag{4}$$

By adding all the Lyapunov exponents in (3), we get their sum as

$$L_1 + L_2 + L_3 + L_4 = -28.2250 < 0 \tag{5}$$

This shows that the new hyperchaotic system (1) is dissipative.

The equilibrium points of the new hyperchaotic system (1) are found by solving the following system of equations:

$$a(x-y) - yz + w = 0 \tag{6a}$$

$$xz - by + w = 0 \tag{6b}$$

$$xy - cz = 0 \tag{6c}$$

$$-x = 0 \tag{6d}$$

From (6d), we get $x = 0$. If we substitute this in (6c), we get $z = 0$.

Then the equations (6a) and (6b) simplify as follows:

$$-ay + w = 0 \tag{7a}$$

$$-by + w = 0 \tag{7b}$$

When $a \neq b$, the system (7) does not admit any solution. For the hyperchaotic case (2), the parameter values are $(a,b,c) = (7,30,10)$.

Thus, we conclude that the new hyperchaotic system (1) does not possess any equilibrium point. Hence, it has hidden attractor [41-45].

Figures 1-4 show the 2-D phase plots of the new system (1). It is seen that the new system (1) exhibits hidden *four-scroll* attractor.

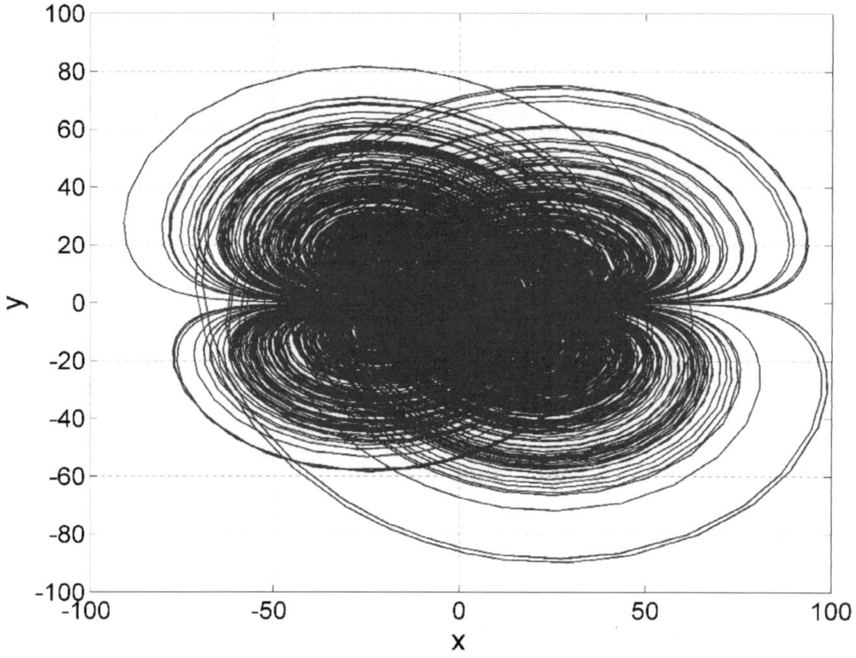

Figure 1. Two-dimensional phase plot of the new hyperchaotic four-scroll system (1) in the $(x_1, x_2)-$plane with $X_0 = (0.1, 0.1, 0.1, 0.1)$ and $(a, b, c) = (7, 30, 10)$

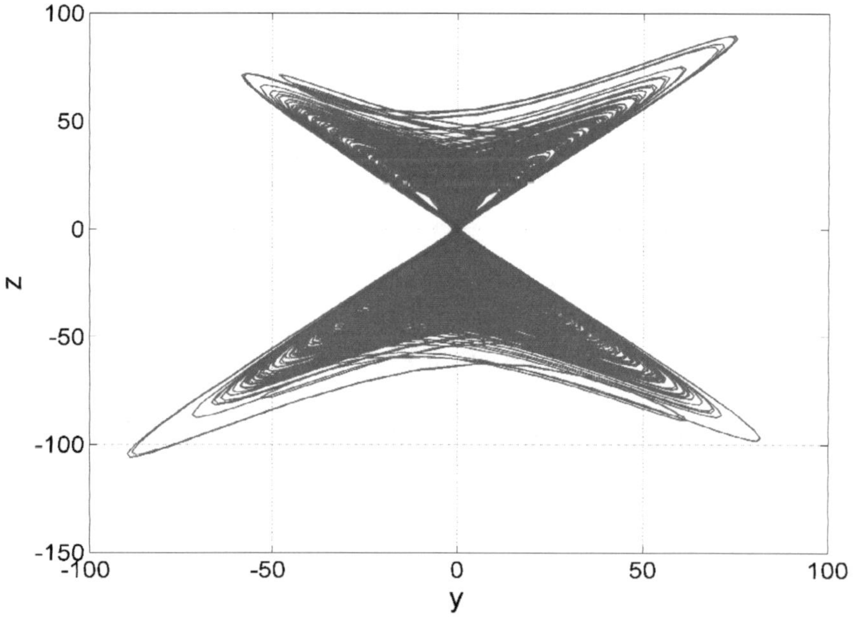

Figure 2. Two-dimensional phase plot of the new hyperchaotic four-scroll system (1) in the $(x_2, x_3)-$plane with $X_0 = (0.1, 0.1, 0.1, 0.1)$ and $(a, b, c) = (7, 30, 10)$

IOP Conf. Series: Materials Science and Engineering **621** (2019) 012014 doi:10.1088/1757-899X/621/1/012014

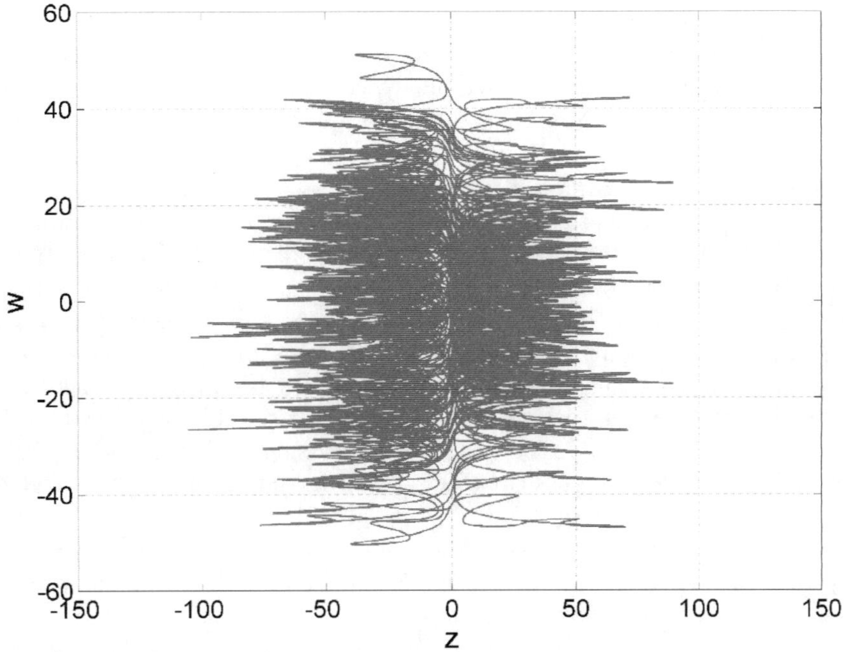

Figure 3. Two-dimensional phase plot of the new hyperchaotic four-scroll system (1) in the $(x_3, x_4)-$plane with $X_0 = (0.1, 0.1, 0.1, 0.1)$ and $(a, b, c) = (7, 30, 10)$

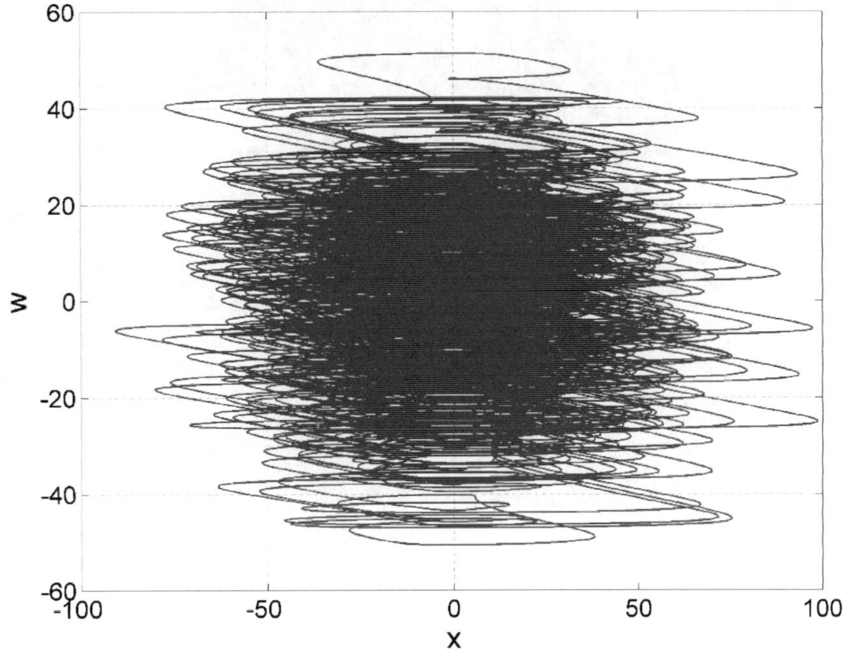

Figure 4. Two-dimensional phase plot of the new hyperchaotic four-scroll system (1) in the $(x_1, x_4)-$plane with $X_0 = (0.1, 0.1, 0.1, 0.1)$ and $(a, b, c) = (7, 30, 10)$

IOP Conf. Series: Materials Science and Engineering **621** (2019) 012014 doi:10.1088/1757-899X/621/1/012014

3. Bifurcation analysis of the new hyperchaotic system

3.1 Route to four-scroll hyperchaos

In this section, we show how the route to four-scroll hyperchaos is obtained using bifurcation diagram and phase portraits. This new four-scroll hyperchaotic system is numerically solved using fourth-order Runge-Kutta formula. For each iteration, the time grid is always $\Delta t = 2\times10^{-4}$ seconds and the computations are made using parameters and variables in extended mode. Bifurcation diagrams and Lyapunov exponent graphs are exploited to define the type of behavior leading to four-scroll hyperchaos. Two strategies are exploited to obtain bifurcation diagram (i.e. upward and backward continuation). The Lyapunov exponent is calculated using the method described by Wolf et al. [46]. In Figure 5, the bifurcation diagram depicts plots of local maxima of the variable x in terms of the control parameter a in the range $0.1 \le a \le 20$. The spectrum of Lyapunov exponents (See Figure 5b) agrees with bifurcation sequence of Figure 5a. From Figure 5a, there are two routes leading to four-scroll hyperchaos. The first route is given by the following sequence of bifurcation when smoothly decreased the control bifurcation parameter a :period-1 $\rightarrow$torus $\rightarrow$four-scroll hyperchaos (See Figure 6). The second route is shown by period-doubling sequence (See Figure 7). Figures 6 and 7 represent 2D projections of phase portraits showing transition to four-scroll hyperchaos as observed in the bifurcation diagram of Figure 5. To better clarify that the new hyperchaotic system develop period-doubling scenario, a bifurcation diagram and corresponding Lyapunov exponents are plotted for varying a in the range $0.2 \le a \le 0.5$ for $b = 30$ and $c = 8.6$ with the same initial condition as fixed in Figure 5. The result is shown in Figure 8.

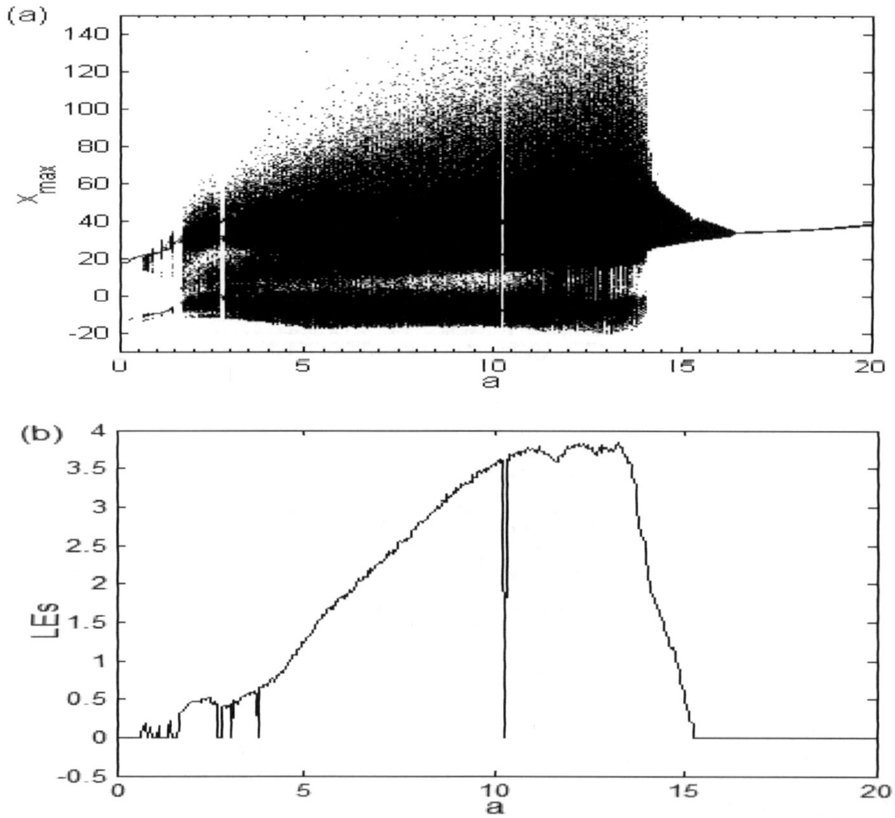

Figure 5. Bifurcation diagram (a) when increasing parameter a , showing local maxima of the coordinate x versus a and the corresponding graph (b) of Lyapunov exponents plotted in the range $0.1 \le a \le 20$. Parameters are fixed as: $b = 30$, $c = 10$.

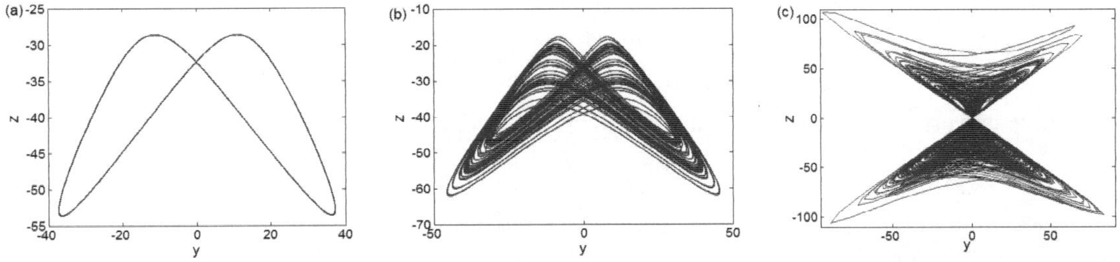

Figure 6. 2D phase portraits of the hyperchaotic system obtained for some specific values of the control parameter a: (a) periodic orbit for $a = 18$, (b) torus state for $a = 15$ and (c) hyperchaos four-wing attractor for $a = 7$. The rest of parameters are the same as in Fig.1.

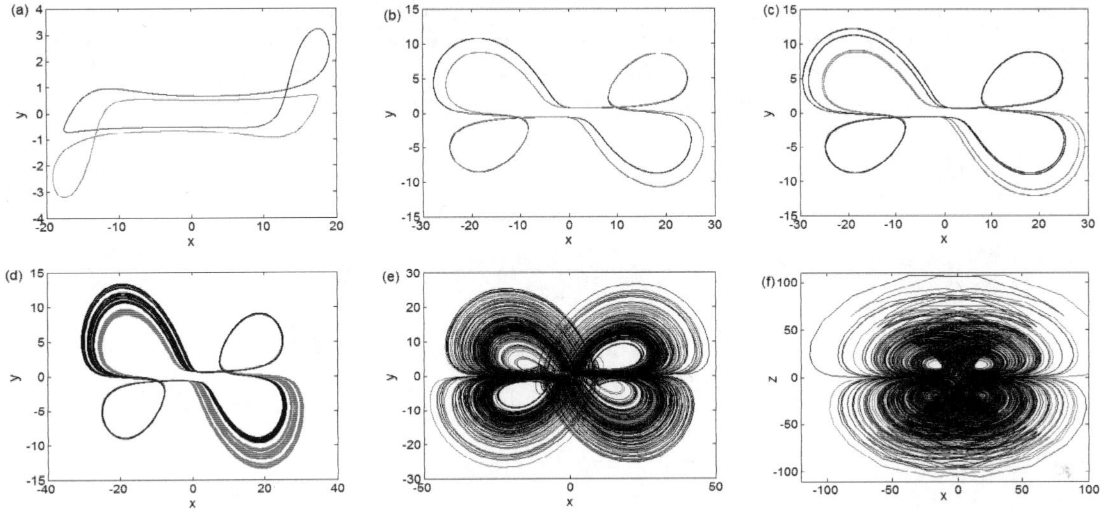

Figure 7. Phase portrait of the new hyperchaotic system showing period-doubling route to hyperchaos four-wing attractor when varied parameter a in the interval $1.2 \le a \le 7$. (a) period-1 limit cycle for $a = 1.25$, (b) period-2 limit cycle for $a = 1.3$, (d) symmetric chaotic attractor for $a = 1.35$, (e) asymmetric chaotic four-wing attractor for $a = 2$, and (f) hyperchaotic four-wing attractor for $a = 7$ whit symmetric initial condition $(\pm 0.7, \pm 0.1, \pm 0.1, \pm 0.1)$.

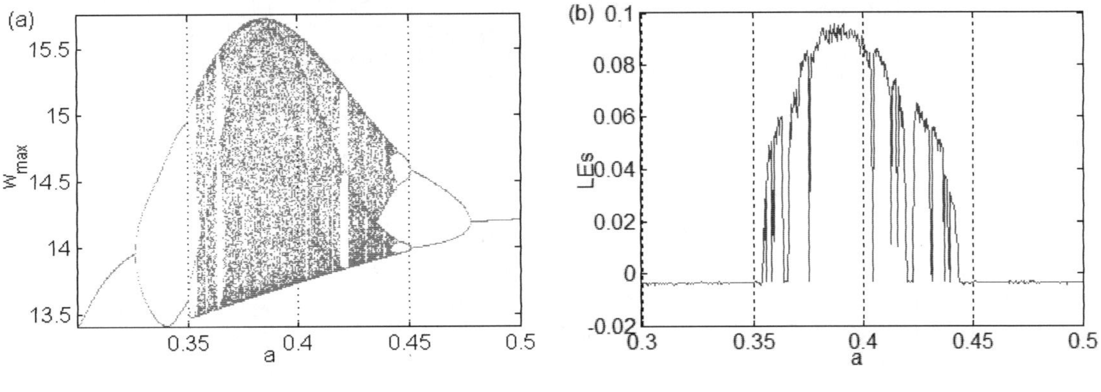

Figure 8. Bifurcation diagrams showing local maxima of the coordinate w of the attractor and corresponding graph of largest Lyapunov exponent in terms of the control parameter a that is varied in tiny steps in the range $0.3 \le a \le 0.5$. This bifurcation diagram show period-doubling route chaotic for increasing parameter a. Initial conditions are $X(0) = (2, 0.1, 0.1, 0.1)$.

3.2 Coexisting bifurcation and multistability

The phenomenon of multiple attractors is observed in the new hyperchaotic system (1) with $b = 30$, $c = 10$ and plotting the bifurcation diagram of the new hyperchaotic system versus the control parameter a (See Figure 9) in a specific range of system parameters. As we see in Figure 9a, a periodic branch (blue diagram) coexists with period-doubling branch (red diagram) while Figure 9b shows bistable property of the model for varying a in the interval $1.2 \le a \le 1.36$. This graph represents an enlargement of the bifurcation diagram depicted in Figure 5. For values of a within this range, the long term behaviour of the system depends solely on its initial state; hence the system develops the interesting phenomenon of multiple coexisting symmetric and asymmetric attractors for appropriate choices of initial conditions [47-50]. For illustration of these coexisting symmetric or asymmetric attractors, we have chosen different initial conditions to summarize each coexisting attractors observed on the bifurcation diagram of Figure 9a. In Figure 10, the possibility of two or three different asymmetric or symmetric periodic and chaotic attractors can be observed when increased the value of parameter a from $a = 1.2$ to $a = 1.8$ (See Figure 10).

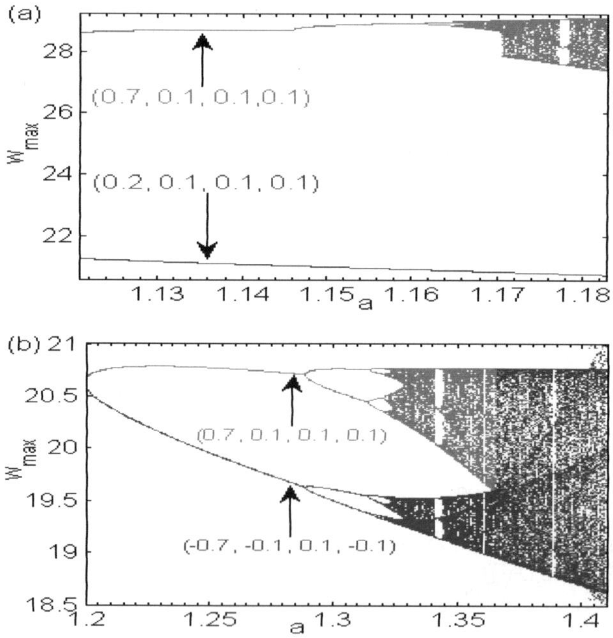

Figure 9. Enlargement of the bifurcation diagrams of Fig.1 showing local maxima of the coordinate w of the attractor in terms of the control parameter a that is varied in tiny steps in the range (a) $1.12 \le a \le 1.18$ and (b) in the range $1.2 \le a \le 1.14$. Theses diagram corresponds for increasing a. Initial conditions are clearly indicated in each graph. In Fig.6a, two or three different coexisting asymmetric and symmetric attractors can be obtained while in Fig.6b show bistability.

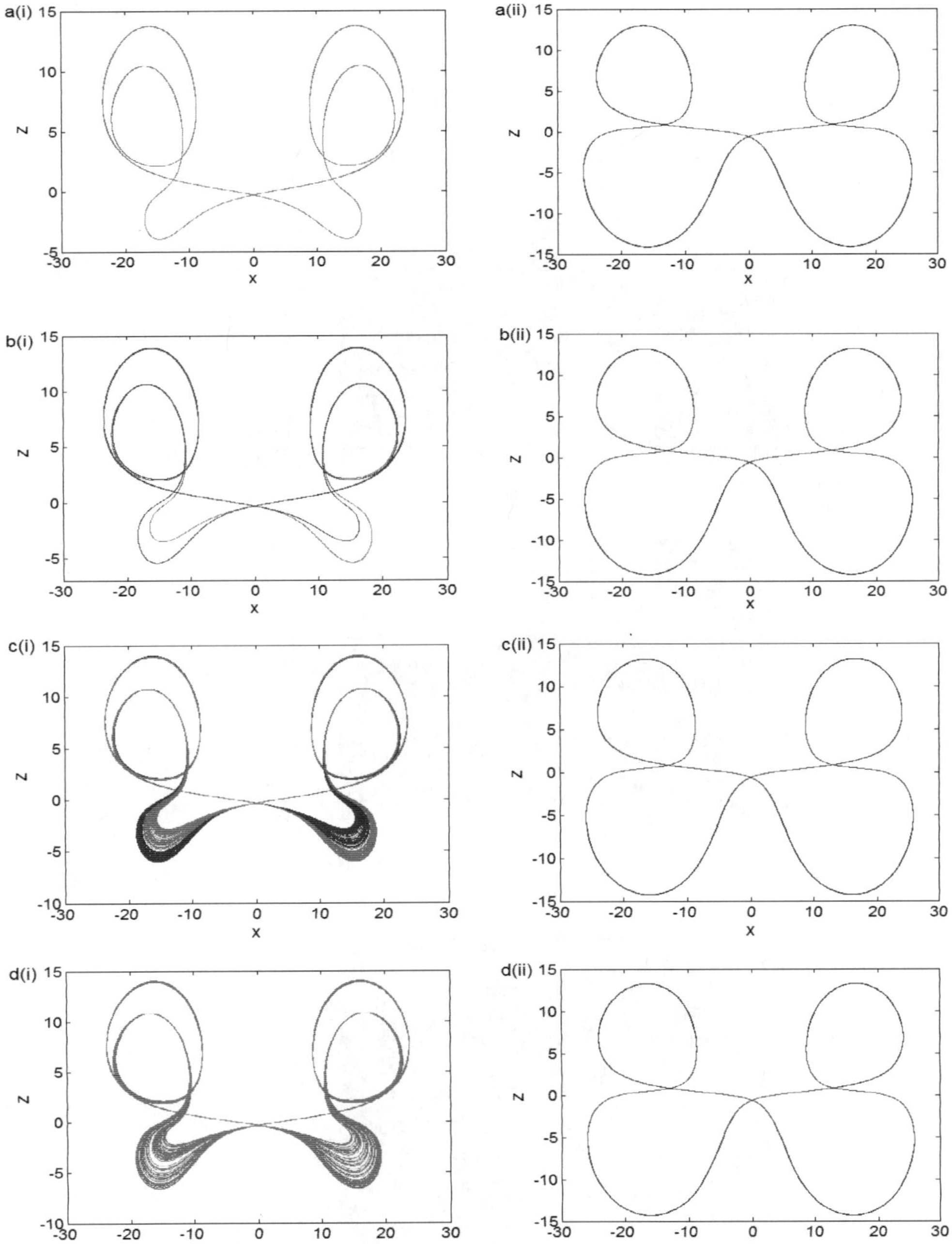

Figure 10. Numerical result showing several coexisting asymmetric or symmetric attractors observe in the bifurcation of Fig. 2a and for different values of a . (a) two asymmetric periodic attractors for $d = 1.13$ with $(0.2, 0.1, 0.1, 0.1)$, and $(0.7, 0.1, 0.1, 0.1)$;(b) two symmetric periodic attractors and one asymmetric periodic attractors for $a = 1.16$ with $(\pm 0.2, 0.1, 0.1, 0.1)$, $(0.7, 0.1, 0.1, 0.1)$;(c) two symmetric chaotic attractors and one asymmetric periodic attractors for $a = 1.17$ with

$(\pm 0.2, 0.1, 0.1, 0.1)$, $(0.7, 0.1, 0.1, 0.1)$;(d) one asymmetric chaotic attractor and one asymmetric periodic attractors for $a = 1.18$ with $(0.2, 0.1, 0.1, 0.1)$, $(0.1, 0.1, 0.1, 0.1)$.

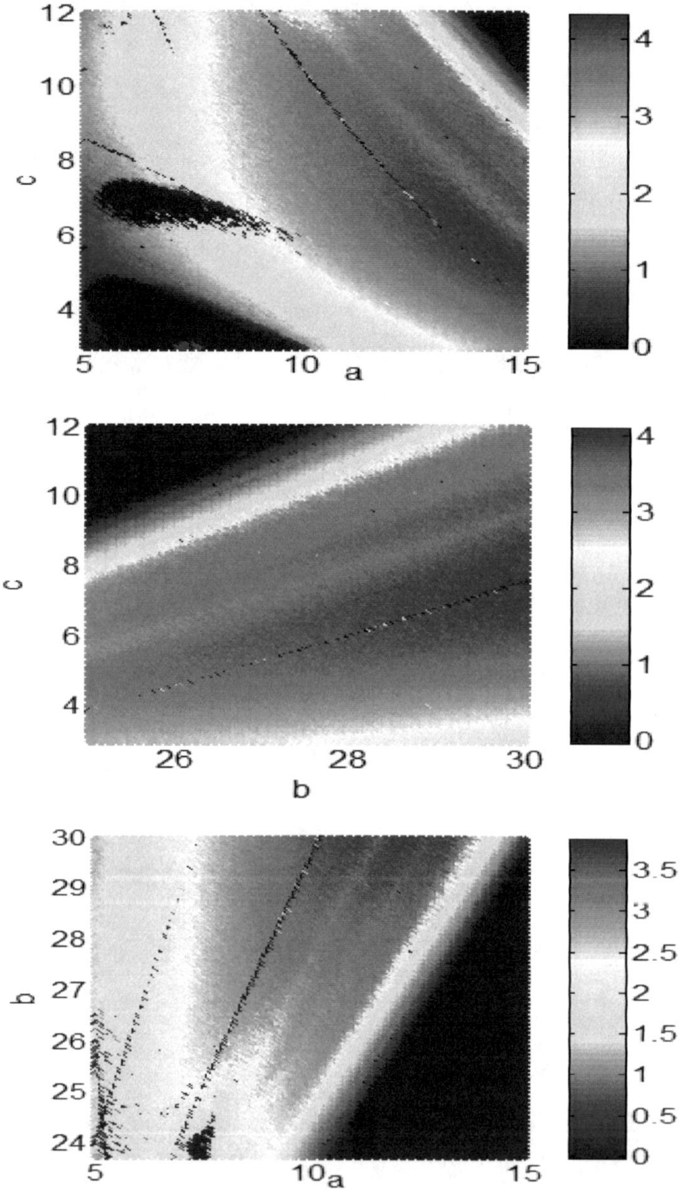

Figure 11. Two-parameter Lyapunov exponent diagrams showing global dynamics of different parameters of the new hyperchaotic system in (a-c), (b-c), and (a-b) planes with same initial conditions $X(0) = (0.1, 0.1, 0.1, 0.1)$. These diagrams are interesting to visualize chaotic or periodic and hyperchaotic domains of new hyperchaotic system.

IOP Conf. Series: Materials Science and Engineering **621** (2019) 012014 doi:10.1088/1757-899X/621/1/012014

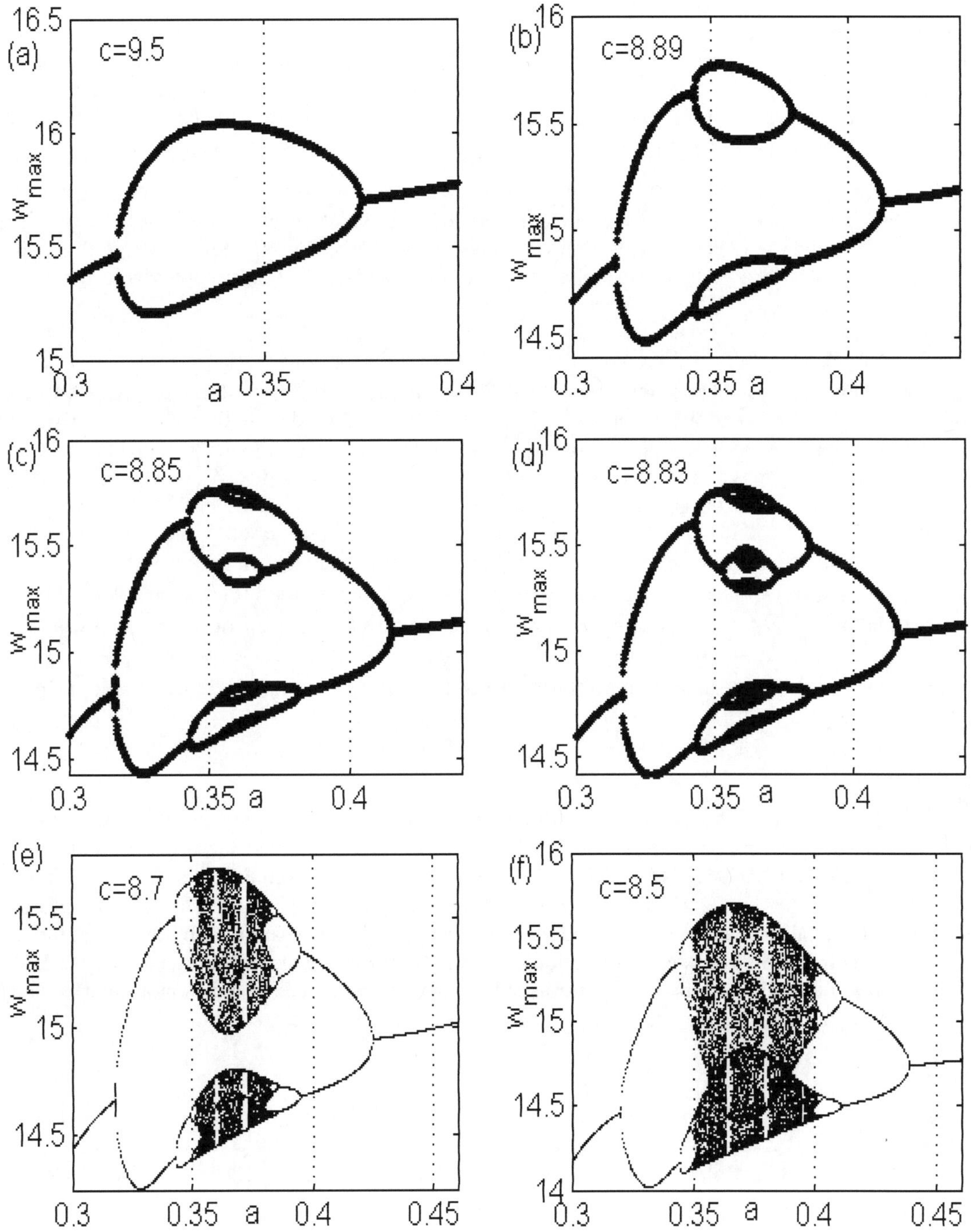

Figure 12. Bifurcation diagrams showing local maxima of the coordinate w of the attractor in Poincaré cross section in terms of the control parameter a showing reversals of period-doubling: periode-2 bubble for $c = 9.5$, periode-4 bubble for $c = 8.89$, periode-8 bubble for $c = 8.85$, chaotic bubble for $c = 8.83$, full chaotic bubble for: $c = 8.7$,and $c = 8.5$. The creation of period doubling follow by their destruction can be clearly identified computed with $b = 30$ and initial state $X(0) = (2, 0.1, 0.1, 0.1))$

3.3 Two parameter Lyapunov exponents

Dynamics of the new hyperchaotic system (1) are studied under the variation of two bifurcation control parameters based on the Lyapunov exponent [51-53]. This study helps to gain a general idea of the dynamic behaviour of the system when two control parameters are varied simultaneously. Figure 11 shows two parameter Lyapunov exponents in (a, b)-plane, (a, c)-plane, and (b, c)-plane with initial conditions $X(0) = (0.1, 0.1, 0.1, 0.1)$ to have more insight on different dynamic behavior. Remark that for each Lyapunov exponent diagrams, the colour changes with the values of the Lyapunov exponents. The graph shows high value of Lyapunov exponent (red domain) which means hyperchaotic zones. The rest of domains are chaotic or periodic oscillation zones. Two-parameter exponent study in this paper has also study in several chaotic and hyperchaotic system to summarize their complex dynamics [51-53].

3.4 Antimonotonicity

Another interesting behavior of the new four-scroll hyperchaotic system investigated in this paper is antimonotonicity (i.e. the creation of period doubling follow by their destruction via reverse period doubling bifurcations). It has been demonstrated in literature that reverse period-doubling occur in nonlinear dynamical system with the variation of two parameters of the system [54]. Reversals of period doubling cascades have been observed in various non-linear physical systems both numerically and experimentally [55-58]. To demonstrate the phenomenon of antimonotonicity in four-scroll hyperchaotic system (1), we have computed some parametric bifurcations of the system using a as control parameter for some discrete values of c. Sample results are provided in Figure 12 where it can be observed that for $c = 9.5$ a period-2 bubble is observed and each branch develops two branches leading to a stable period-4 bubble for $c = 8.89$. Similarly a period-8 bubble is obtained for $c = 8.85$. As the control parameter c decreases, more bubbles are created until an infinitely tree (chaos) finally occurs when $c = 8.5$.

4. Conclusions

In this paper, we reported our results of a new four-dimensional hyperchaotic system with a four-scroll attractor and discusses its dynamic properties such as Lyapunov exponents, phase portraits, Kaplan-Yorke dimension and equilibrium points. Our calculations of the new hyperchaotic system demonstrated that the new hyperchaotic system has no equilibrium point and hence it exhibits hidden attractor. A detailed bifurcation analysis has been presented for the new hyperchaotic system with four-scroll hidden attractor. Specifically, we discussed bifurcation analysis such as route to four-scroll hyperchaos, coexisting bifurcation, multistability, two parameter Lyapunov exponents and antimonotonicity. Many simulation plots using MATLAB were shown to describe our results in detail.

References

[1] Azar A T and Vaidyanathan S 2016 Advances in Chaos Theory and Intelligent Control (Berlin: Springer)

[2] Vaidyanathan S and Volos C 2016 Advances and Applications in Chaotic Systems (Berin: Springer)

[3] Vaidyanathan S and Volos C 2016 Advances and Applications in Nonlinear Control Systems (Berlin: Springer)

[4] Pham V T, Vaidyanathan S, Volos C, Kapitaniak T 2018 Nonlinear Dynamical Systems: Self-Excited and Hidden Attractors (Berlin: Springer)

[5] Pham V T, Volos C K and Vaidyanathan S 2015 *Studies in Computational Intelligence* **581** 59-72

[6] Vaidyanathan S, Volos C K and Pham V T 2015 *Studies in Computational Intelligence* **576** 571-590

[7] Vaidyanathan S and Rasappan S 2011 *Communications in Computer and Information Science* **131** 585-593
[8] Vaidyanathan S 2012 *Lecture Notes of the Institute for Computer Sciences, Social-Informatics and Telecommunications Engineering* **85** 124-133 2012
[9] Vaidyanathan S 2015 *Int. J. Modelling, Identification and Control* **23** 380-392
[10] Vaidyanathan S 2015 *Kyunjgpook Mathematical Journal* **55** 563-586
[11] Rasappan S and Vaidyanathan S 2012 *Far East J. Mathematical Sciences* **67** 265-287
[12] Vaidyanathan S, Volos C K, Rajagopal K, Kyprianidis I M and Stouboulos I N 2015 *J. Engineering Science and Technology Review* **8** 74-82
[13] Vaidyanathan S 2015 *Int. J. ChemTech Research* **8** 209-221
[14] Vaidyanathan S 2015 *Int. J. ChemTech Research* **8** 146-158
[15] Vaidyanathan S 2015 *Int. J. ChemTech Research* **8** 159-171
[16] Vaidyanathan S, Sambas A, Mamat M and Sanjaya W S M 2017 *Archives of Control Sciences* **27** 541-54
[17] Mansour S M B, Sundarapandian V and Naceur S M 2016 *International J. Control Theory and Applications* **9** 37-54
[18] Akgul A, Moroz I, Pehlivan I and Vaidyanathan S 2016 *Optik* **127** 5491-99
[19] Zhang Y 2018 *Information Sciences* **450** 361-77
[20] Gunasekaran G and Venkatesan M 2018 *Pertanika J. Science and Technology* **26** 599-614
[21] Pham V T, Jafari S, Volos C, Giakoumis A, Vaidyanathan S and Kapitaniak T 2016 *IEEE Trans. Circuits and Systems II: Express Briefs* **63** 878-82
[22] Volos C, Maaita J O, Vaidyanathan S, Pham V T, Stouboulos I and Kyprianidis I 2017 *IEEE Trans. Circuits and Systems II: Express Briefs* **64** 339-43
[23] Vaidyanathan S, Pham V T, Volos C and Sambas A 2018 *Studies in Systems, Decision and Control* **133** 345-64
[24] Sambas A, Vaidyanathan S, Mamat M and Mada Sanjaya W S 2018 *Studies in Systems, Decision and Control* **133** 365-73
[25] Vaidyanathan S, Sambas A, Mamat M and Sanjaya W S M 2017 *Int. J. Modelling, Identification and Control* 28 153-66
[26] Sambas A, Vaidyanathan S, Mamat M, Sanjaya W S M and Prastio R P 2016 *Int. J. Control Theory and Applications* **9** 141-49
[27] Rössler O E 1979 *Physics Letters A* 71 155-57
[28] Vafamand N, Khorshidi S and Khayatian A 2018 *Chaos Solitons Fractals* **112** 116-24
[29] Zaher A A 2009 *Chaos Solitons & Fractals* **42** 2804-14
[30] He J, Cai J and Lin J 2016 *Optik* **127** 2502-08
[31] Vaidyanathan S 2017 *Studies in Computational Intelligence* **701** 393-423
[32] Vaidyanathan S, Pham V T and Volos C *Studies in Computational Intelligence* **701** 101-30
[33] Pham V T, Vaidyanathan S, Volos C, Wang X and Hoang D V 2017 *Studies in Computational Intelligence* **701** 59-80
[34] Wu X, Bai C and Kan H 2014 *Communications in Nonlinear Science and Numerical Simulation* **19** 1884-97
[35] Li C, Liao X and Wong K 2005 *Chaos Solitons & Fractals* **23** 183-93
[36] Boriga R, Dascalescu A C and Priescu I 2014 *Signal Processing* **29** 887-901
[37] Zhou N, Pan S, Cheng S and Zhou Z 2016 *Optics and Laser Technology* **82** 121-33
[38] Jajarmi A, Hajipour M and Baleanu D 2017 *Chaos Solitons & Fractals* **99** 285-96
[39] Vaidyanathan S and Rajagopal K 2017 *Int. J. Simulation and Process Modelling* **12** 165-78
[40] Vaidyanathan S, Akgul A, Kacar S and Cavusoglu U 2018 *European Physial Journal Plus* **133** 46
[41] Leonov G A 2016 *Doklady Physics* **61** 340-42
[42] Dudkowski D, Jafari S, Kapitaniak T, Kuznetsov N V, Leonov G A and Prasad A 2016 *Physics Reports* **637** 1-50

IOP Conf. Series: Materials Science and Engineering **621** (2019) 012014 doi:10.1088/1757-899X/621/1/012014

[43] Leonov G A 2017 *Communications on Pure and Applied Analysis* **16** 2253-67
[44] Vaidyanathan S, Pham V T, Volos C and Sambas A 2018 *Studies in Systems, Decision and Control* **133** 345-64
[45] Vaidyanathan S, Pham V T and Volos C K 2015 *European Physical Journal: Special Topics* **224** 1575-92
[46] Wolf A, Swift J B, Swinney H L and Vastano J A 1985 *Physica D* **16** 285-317
[47] Li C, Hu W, Sprott J C and Wang X 2015 *European Physical Journal: Special Topics* **224** 1493–1506
[48] Li C and Sprott J C 2014 *Int. J. of Bifurcation and Chaos* **24** 1450034
[49] Li C, Sprott J C and Xing H 2017 *Nonlinear Dyn.* **87** 1351-58
[50] Leutcho G D, Kengne J and Kamdjeu Kengne L 2018a *Chaos, Solitons & Fractals* **107** 67-87
[51] Stegemann C, Albuquerque H A, Rubinger R M and Rech P C 2011 *Chaos: Int. J Nonlinear Sci.* **21** 033105
[52] Rech P C 2017 *Eur Phys J B* **90** 251
[53] Fozin Fonzin T, Srinivasan K, Kengne J and Pelap F B 2018 *AEU Int. J. Electron. Commun.* **90** 110-112
[54] Kyprianidis I, Stouboulos I, Haralabidis P and Bountis T 2000 *Int. J. Bifurcation and Chaos* **10** 1903-1915
[55] Leutcho G D, Kengne J and Kengne R 2018 *Int. J. Dynam. Control*, Doi: 10.1007/s40435-018-0438-7
[56] Leutcho G D and Kengne J 2018 *Chaos, Solitons & Fractals* **113** 275-93
[57] Nguomkam Negou A and Kengne J 2018 *Int. J. Electronics and Communications* **90** 1-19
[58] Kocarev L, Halle K, Eckert K and Chua L 1993 *Int. J. Bifurcation and Chaos* **3** 1051-55

Modeling the impacts of constant price GDP and population on CO2 emissions using Cobb-Douglas model and ant colony optimization algorithm

Sukono[1], Betty Subartini[1], Priscila Thalia, Sudradjat Supian[1], Eman Lesmana[1], Ruly Budiono[2], Hafizan Juahir[3]

[1]Department of Mathematics, Faculty of Mathematics and Natural Sciences, Universitas Padjadjaran, Indonesia

[2]Department of Biology, Faculty of Mathematics and Natural Sciences, Universitas Padjadjaran, Indonesia

[3]East Coast environmental Research Institute (ESERI), Universiti Sultan Zainal Abidin, Malaysia

Email: sukono@unpad.ac.id; betty.subartini@unpad.ac.id; friscilathalia31@gmail.com; sudradjat@unpad.ac.id; man.msie@gmail.com; budiono.ruly@gmail.com; hafizanjuahir@unisza.edu.my

*corresponding author: sukono@unpad.ac.id

Abstract. The per capita Gross Domestic Product (GDP) measures a country's economic growth. Increasing GDP is a dream of all countries, but generally, GDP increases often have a negative impact with increasing CO_2 emissions. This paper intends to model the impact of GDP growth based on constant prices and the population in increasing CO_2 emissions in Indonesia. Modeling is done by using Cobb-Douglas model production function, where parameter estimation is done by using ant colony optimization algorithm. Furthermore, model estimators are used for forecasting CO_2 emission concentrations. The results of the analysis show that the impact of GDP based on constant prices and population significantly follows the Cobb-Douglas model of production, with the coefficient of elasticity is 0.819405999 and 0.834930855, respectively. The value of determination was obtained at 97.4%, indicating that the correlation between GDP at constant prices and population with increasing CO_2 emissions in air is very strong. Estimator model obtained has a level of accuracy for forecasting is 0.98478981 or 98.4798981%. Thus, the model estimator obtained is able to describe the actual data pattern.

Keywords: GDP, Population, CO_2 emissions, Cobb-Douglas model, ant colony optimization.

1. Introduction

Economic development is an activity to promote economic growth for every country. One of the economic growths is measured by the amount of Gross Domestic Product (GDP) per capita. However, high GDP growth is often followed by declining environmental quality. GDP growth is not the only one to be a successful measure of development of a country [1; 2]. Keeping nature and the environment alive is also a factor of development success. GDP growth along with population growth also has a negative effect, and significantly affects the quality of the environment as indicated by increased CO_2 emissions [7; 9]. Any 1 percent increase in GDP and population, will lead to a decrease in environmental quality by 9.11 percent. It shows that GDP growth and population growth will lead to a decrease in the level of environmental quality in the form of rising CO_2 emissions [8; 10].

Thus, GDP growth and population growth have an impact on increasing CO_2 emissions in the air [17; 19]. How patterns of GDP growth and population growth together contribute to increasing CO_2 emissions, this becomes an important issue for a study. Previous researchers, among them, have done many studies are: Laceheb et al. [6] reviewed the economic growth and CO_2 emissions in Algeria in the period 1971-2009. The study was conducted using an integrated autoregressive lag model, where the data were extracted from the World Bank Indicators. The results show that over the long term, it was revealed that income and population growth significantly affected CO_2 emissions. The results of the study also show that only the population showed increased CO_2 emissions from liquid fuel consumption. Yang et al. [15], investigating the interaction of economic growth and pollution in the industrialization process in Zhejiang, time series data from three types of pollution indices from 1981 to 2006. The investigation is done by using cointegration test and granger causality test. The test results show that GDP per capita results in increased emissions of industrial waste pollution except solid waste. Conversely, it is not true that industrial waste pollution increases can lead to economic growth. Conrad & Cassar [3], Sasana & Putri [13], and Zhao & Wang [18] have also conducted similar studies.

Salami [11], in his study proposed a hybrid algorithm to find solutions to optimization problems using ant colony and genetic programming. An evolutionary process of ant colony optimization algorithms adapts genetic operations to improve ant motion to achieve optimal solutions. The proposed algorithm integrates with the optimal solution, by collecting the most effective sub-solutions. Samsami [12] conducted a study on the application of Ant Colony Optimization (ACO) to predict CO_2 emissions in Iran, based on socio-economic indicators. Forms of linear and non-linear equations were developed to predict CO_2 emissions using ACO. The results obtained from the study provide useful insights into energy systems and CO2 emission control modeling. Other studies that have done the same are Duan & Yong [4] and Yu & Wang [16].

Therefore, based on the exposure of previous researchers, in this paper the research on the modeling of the impact of GDP at constant price and population on CO_2 emission is done by using Cobb-Douglas model and particle swarm optimization algorithm. The research objects are GDP growth data and population, and CO_2 emissions in Indonesia in period 1967-2014. The goal is to derive a significant model estimate, and can be used for forecasting the development of CO_2 emissions with minimum errors or high accuracy.

2. Methodology

This methodology discusses some of the models and algorithms used in the study, which include Cobb-Douglas production function, model parameter estimation using ant colony optimization (ACO), model significance test, and forecasting (prediction).

2.1. Cobb-Douglas model

Referring to Felipe & Adams [5] and Omri [10], the Cobb-Douglas production function is an equation that contains two or more variables, a dependent variable and several independent variables. Cobb-Douglas multiplicative error term term production function is expressed by the following equation:

$$G_t = \phi X_{1t}{}^{\theta_1} X_{2t}{}^{\theta_2} e^{\varepsilon_t}, \tag{1}$$

where G_t is to declare an output variable in the form of CO2 emissions; X_{1t} is to state the input variable in the form of GDP at constant price per capita; X_{2t} is to declare the input variable the number of population; ϕ, θ_1, and θ_2 is to declare the Cobb-Douglas model parameters; and ε_t is exponential of residual.

Referring to Felipe and Adams [5], the coefficient of elasticity of production ρ is to state the percentage change in output, divided by the percentage of input changes. The coefficient of production elasticity is the ratio of the relative change of output produced, to the relative changes in the number of inputs that affect. The coefficient of output elasticity of GDP at constant prices input, ρ_1 calculated using the equation as follows:

$$\rho_1 = \frac{\%\Delta G_t}{\%\Delta X_{1t}} = \theta_1 .$$

The output elasticity coefficient of GDP at constant price input can also be determined by using coefficient parameters θ_1 of the Cobb-Douglas production function.

While the output elasticity coefficient of the input population, ρ_2 can be calculated using the equation as follows:

$$\rho_2 = \frac{\%\Delta G_t}{\%\Delta X_{2t}} = \theta_2 .$$

The ouput elasticity coefficient of the input populationcan also be calculated by using parameter coefficient θ_2 of the Cobb-Douglass production fucntion.

Referring to Felipe and Adams [5], the sum of the production elasticity coefficients $\theta_1 + \theta_2$ explain the size of the venture scale or often called return to scale. There are three forms of return to scale, as given below:

- When $\theta_1 + \theta_2 = 1$, the Cobb-Douglas production function shows a constant return to scale, meaning that the increase in input will be followed by an equivalent proportional output increase.
- When $\theta_1 + \theta_2 < 1$, Cobb-Dougal production function shows the scale with decreasing return (decreasing return to scale), meaning that the proportional increase of output is smaller than the proportional addition of input.
- When $\theta_1 + \theta_2 > 1$, Cobb-Douglas production function shows scale with increasing return (increasing return to scale), meaning that proportional addition of output greater than proportional addition of input.

Equation (1), if the left and right segments are taken natural logarithm, then obtained multiple linear regression equation as follows:

$$\ln G_t = \ln \phi + \theta_1 \ln X_{1t} + \theta_2 \ln X_{2t} + \varepsilon_t .$$

Suppose that $Y_t = \ln G_t$; $\theta_0 = \ln \phi$; $Z_{1t} = \ln X_{1t}$; and $Z_{2t} = \ln X_{2t}$, then the equation can be expressed as follows:

$$Y_t = \theta_0 + \theta_1 Z_{1t} + \theta_2 Z_{2t} + \varepsilon_t . \tag{2}$$

The estimate of the multiple linear regression equation (2) is as follows:

$$\hat{Y}_t = \theta_0 + \theta_1 Z_{1t} + \theta_2 Z_{2t} . \tag{3}$$

Equation (2) is a linear regression with parameters θ_0, θ_1, and θ_2 as well as residuals (error) ε_t. Constants θ_0 is intercept, θ_1 and θ_2 is a coefficient parameter of elasticity of Cobb-Douglas production function.

If the estimation is done by using ordinary least square method (OLS, then from multiple linear regression equation (2) it is necessary to establish the optimization model as follows:

$$\text{Minimisasi } \sum \varepsilon_t^2 = \sum (Y_t - \hat{Y}_t)^2 = \sum (Y_t - \theta_0 - \theta_1 Z_{1t} - \theta_2 Z_{2t})^2 . \tag{4}$$

It means that, the estimation is aimed at determining the estimator values of θ_0, θ_1, and θ_2 which can minimize the sum of squares of residuals $\sum \varepsilon_t^2$. The process of minimizing the number of residual squares, in this research is done by using Ant Colony Optimization (ACO) algorithm.

2.2. Ant coloni optimization algorithm

According to Duan & Yong 4], Salami [11], Samsami [12], and Yue & Wang [16], Ant Colony Optimization (ACO) is adopted from the treatment of ant colonies, called ant systems. Naturally, ant colonies are able to find the shortest route on the way from the nest to the source of food and back to the nest. At the time of ants traveling, ants always leave a piece of information called pheromone, where it travels and to mark the route. Pheromone works for communication between ants when creating routes. The road route the ant passes from the nest to where the food can be given as Figure 1. The ACO algorithm has been widely used to solve various optimization cases.

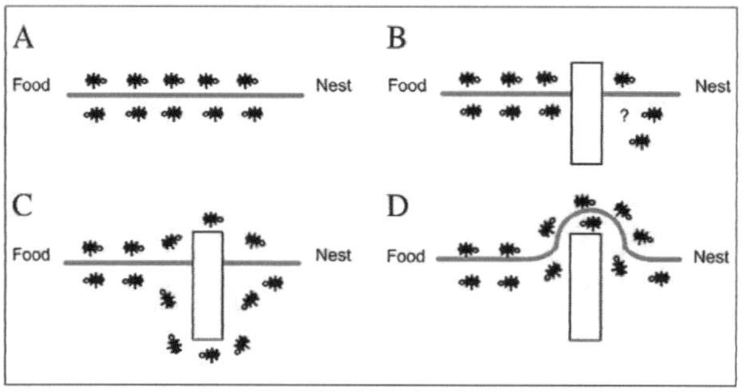

Figure 1. The route of the ant path from the nest to the food place

According to Duan & Yong [4], Salami [11], Samsami [12], and Yue & Wang [16], the workings of the Ant Colony algorithm are described as follows:

1. Initially, ant colonies stroll randomly.
2. When the ant colony finds a different path, such as finding the intersection, the ant colony determines the direction of the path randomly.
3. Some ant colonies choose to walk to the upper route, and some choose to walk down the route.
4. Whenever he finds some ant food back to his colony, while putting a sign with a pheromone trail.
5. Because the path taken down the shorter route, the ant colony passing down first arrives, which assume the velocity of all ant colonies to be the same.
6. The pheromone left by the ant colony on the shorter route smells stronger, than the pheromone on the longer path.
7. Other ant colonies are more interested in following the route down, because the pheromone aroma is stronger.

According to Duan & Yong [4], Salami [11], Samsami [12], and Yue & Wang [16], the ant colony algorithm requires several variables, and the steps to determine the shortest distance.

Stage 1:

a. The parameters required in the ant colony algorithm are as follows:

- Intemity of ant colony traces between places τ_{ij} and delta changes $\Delta\tau_{ij}$.

Intensities τ_{ij} should be initialized before starting the iteration cycle. Change $\Delta\tau_{ij}$ initialized after the completion of one iteration cycle. Change $\Delta\tau_{ij}$ used to determine τ_{ij} on the next iteration cycle.

- The ants colony iteration cycle stays Q
  The iteration cycle Q is a constant used in the equation used to determine the delta of change $\Delta\tau_{ij}$. The iteration cycle value Q determined by the user.
- Ant colony trace intensity control constant α
  Traffic intensity control constant α used in the probability equations of places visited, and serves as control of the intensity of ant colony traces. Value α determined by the user.
- Time visibility controller β
  Time visibility controller β is used in the probability equation of locations visited by ant colonies, and serves as a visibility controller. Value β determined by the user.
- Visibility between places η_{ij}
  Visibility between places η_{ij} used in the probability equation of places visited by ant colonies. Value η_{ij} is the result of $1/d_{ij}$ (the distance of the place).
- The number of ant colonies m
  The number of ant colonies m, is the number of ant colonies that the iteration cycle does in the ant colony algorithm. Value m determined by the user.
- The vaporization constant of ant colony traces ρ
  The vaporization constant of ant colony traces ρ used to determine for τ_{ij} on the next iteration cycle. Value ρ determined by the user.
- Number of maximum iteration cycles $NC_{\max}$
  Number of maximum iteration cycles $NC_{\max}$, is the maximum number of shrimp cycles that will take place. The iteration cycle will stop according to the value $NC_{\max}$ pre-defined by the user.
- Charging the coordinates of the place
  On the filling of the coordinates of the place can be input in accordance with the desired by the user. The more coordinates, the longer the ant colony travels.

b. Initialize first place of each ant colony.

After initialization τ_{ij} done, then m ant colonies are placed in a particular random first place. For parameter values α should be rated between ranges $0 \leq \alpha \leq 1$, it is intended to avoid unlimited accumulation of pheromone on the route. Because the amount of pheromone left behind is unlikely to grow stronger, it gets weaker. For parameter values β should not be given a value of 0, because if given a value of 0 then the results achieved are not optimum. Not optimum here means a condition where the length of the journey achieved is not minimum.

Stage 2:

Filling the first place into the taboo list. The initialization result of the first place of each ant in stage 1, must be filled as the first element in the taboo list. The result of this step is the taboo list for each ant colony with a specific spot index.

Stage 3:

Arrange the route of each ant colony's visit to a number of places. An ant colony that has spread to a number of places, began to travel from the first place of each as a place of origin, and to one of the other places as the destination. Then from the second place, each ant colony continues its journey, choosing one of the places that is not on the taboo list, as the next destination. Ant colony trips continue continuously until all the places one by one visited. If s specifies the index of the order of visits, the place of origin is expressed as $tabu_k(s)$, and other places are declared as $\{N - tabu_k(s)\}$, then to determine the destination, used the probability equation where to visit as follows:

$$P_{ij}^k(s) = \begin{cases} \dfrac{[\tau_{ij}(s)]^{\alpha}[\eta_{ij}]^{\beta}}{\sum\limits_{k' \in \{N - tabu_k(s)\}} [\tau_{ij}(s)]^{\alpha}[\eta_{ij}]^{\beta}}; \text{if } j \in \{N - tabu_k(s)\}. \\ 0; \text{for other } j. \end{cases} \tag{5}$$

where i as the index of place of origin, and j as the destination index.

Stage 4:

a. Calculation of the route length of each ant colony
Long route calculation of closed or L_k each ant colony is performed after all ants complete one iteration cycle. Calculations are based $tabu_k(s)$ each with the following equation:

$$L_k = d_{tabu_k(n), tabu_k(1)} + \sum_{s=1}^{n-1} d_{tabu_k(s), tabu_k(s+1)}, \tag{6}$$

where d_{ij} is the distance between i and j which is calculated using following equation:

$$d_{ij} = \sqrt{(x_i - x_j)^2 (y_i - y_j)^2}. \tag{7}$$

b. Shortest distance search
After L_k each ant colony is calculated, there will be a minimum price of the closed route length of each cycle, or $L_{\min} NC$ and the minimum price of the overall closed line length is $L_{\min}$.

c. Determination of the value of price change of the ant colony footprint intensity between places $\Delta\tau_{ij}$
An ant colony leaves footprints on the route between places it passes. The existence of evaporation and differences in the number of ant colonies that pass, causing the possibility of changes in the price of ant colony footprint intensity between places. The value of the price change in the intensity of the ant colony footprint is determined using the equation:

$$\Delta\tau_{ij} = \sum_{k=1}^{m} \Delta\tau_{ij}^k, \tag{8}$$

where m is the number of ant colonies; and τ_{ij} is the length of each ant colony's route. Where $\Delta\tau_{ij}^k$ is the value of the price change of the ant colony footprint intensity between places, for each ant colony calculated using the equation:

$$\Delta\tau_{ij}^k = \begin{cases} \dfrac{Q}{L_k}; \text{for } (i,j) \in \text{place of origin and destination in tabu}_k, \\ 0; \text{for other } (i,j) \end{cases} \tag{9}$$

where Q is the ant colony iteration cycle of the ants; and L_k is length close tour (lct).

Stage 5:

a. Calculation of intem colony footprint intensity values between places for subsequent iteration cycles. The intensity of the ant colony footprints between places on all routes is likely to change, as there is evaporation, and the differences and the number of ant colonies passing through. For the next iteration cycle, the ant colony passing through the route the intensity value has changed. The intensity of the ant colony footprint between places for an iterative cycle is then calculated using the equation:

$$\tau_{ij} = (1-\rho)\times\tau_{ij} + \Delta\tau_{ij}\,. \tag{10}$$

b. Rearranging the change in the intensity of ant colony footprints between places. For the next iteration cycle the change in the intensity value of ant colony traces between places needs to be reset to have a value equal to zero.

Stage 6:

Dismiss the taboo list and repeat step 2 if needed. Taboo lists need to be emptied to be filled again with a new place sequence in the next iteration cycle. If the maximum number of iteration cycles has not been reached, the algorithm is repeated from the taboos filling stage, with the ant colony footprint intensity parameter values between the updated places.

2.3. The significance test of model parameters

The significance test of model parameters is done by using simultaneous test (statistical test of F), test individually (statistical test of t), assumption of residual normality, and determination of efficiency. In this study, the significance test of model parameters was done using Minitab 16 assistance.

- Statistic test of F

This test is conducted with the aim to know that all parameter estimators simultaneously significantly affect the dependent variable. The parameter significance test simultaneously the hypothesis used is: H_0: $\theta_0 = \theta_1 = \theta_2 = 0$, meaning that free variables do not affect significantly to the dependent variable. While the alternative hypothesis is H_1: $\exists\ \theta_0 \neq \theta_1 \neq \theta_2 \neq 0$, meaning that at least there are independent variables that have a significant effect on the dependent variable. The value of count statistics F_{Stat} is determined using the equation [14; 5]:

$$F_{Stat} = \frac{\text{MS}_{\text{Reg}}}{\text{MS}_{\text{Error}}} = \frac{\dfrac{\sum_{t=1}^{n}(\hat{Y}_t - \bar{Y})^2}{k-1}}{\dfrac{\sum_{t=1}^{n}(Y_t - \hat{Y}_t)^2}{n-k}}\,. \tag{11}$$

Next, the calculated statistical value F_{Stat} is compared with statistical critical values $F_{(\alpha,df)}$ with a significant level α and degree of freedom $df = (k-1, n-k)$. Where n is number of samples, and k is the number of parameters. The test criterion is to reject the hypothesis H_0 and to accept the hypothesis H_1, if $F_{Stat} > F_{(\alpha,df)}$. This means that the independent variables affect the independent variable. Instead, to accept the hypothesis H_0 and to reject the hypothesis H_1, if $F_{Stat} \leq F_{(\alpha,df)}$. Meaning that simultaneously independent variable does not affect the dependent variable.

- Statistic test of t

Individual significance test parameters, conducted with the aim to determine whether each parameter estimator significantly contributes to affect the dependent variable. The significance test of individual parameters is done with the hypothesis used is H_0: $\theta_i = 0$, $(i = 0,1,2)$, means that the independent variable does not significantly affect the dependent variable. While the alternative hypothesis is H_1: $\theta_i \neq 0$, $(i = 0,1,2)$, meaning that that the independent variables significantly influence the dependent variable. Value statistics t_{Stat} is determined using the equation [14; 5]:

$$t_{stat} = \frac{\theta_i}{SE(\theta_i)}. \tag{12}$$

Next, the calculated statistical value t_{Stat} is compared with statistical critical values $t_{(\frac{\alpha}{2},df)}$ with a significant level α and degree of freedom $df = (n-k)$. Where n is number of samples, and k is the number of parameters. The testing criterion is to reject the hypothesis H_0 and to accept the hypothesis H_1, if $|t_{Stat}| > |t_{(\frac{\alpha}{2},df)}|$. This means that each independent variable significantly affects the dependent variable. Instead accepting the hypothesis H_0 and rejecting the hypothesis H_1, if $|t_{Stat}| \le |t_{(\frac{\alpha}{2},df)}|$. This means that each independent variable has no significant effect on the dependent variable.

- Residual normality test

The normality assumption test is performed to determine that the residual data follows the normal distribution. Normality assumption test in this paper is done by using Kolmogorov-Smirnov (KS) statistic. In this test, the hypothesis used is H_0: the data is normally distributed, with the alternative hypothesis being H_1: data is not normally distributed. Test assumption of residual normality is done by determining standard deviation by using equation:

$$S_{\varepsilon_i} = \sqrt{\frac{\sum_{i=1}^{n}(\varepsilon_i - \bar{\varepsilon})^2}{n-1}}. \tag{13}$$

Then, transformed value ε_i into z_i using equations $z_i = (\varepsilon_i - \bar{\varepsilon})/S\varepsilon_i$. While the value of probability $P(z_i)$ using standard normal distribution tables. For probability $S(z_i)$ determined using the equation $S(z_i) = randl(z_i)/n$. Next, calculated the values of absolute difference $|S(z_i) - P(z_i)|$. The statistical value of Kolmogorov-Smirnov KS_{Stat} determined using the expression [14]:

$$KS_{Stat} = \max\{|S(z_i) - P(z_i)|\}. \tag{14}$$

Determine the critical value of statistics $KS_{(\alpha,n-1)}$, at a significant level $\alpha = 0.05$. The criteria of the test is to reject the hypothesis H_0 if statistics $KS_{Stat} > KS_{(\alpha,n-1)}$.

- Coefficient of determination value

Determination coefficient value is used to measure correlation strength between all independent variables with non-free variable. Value of coefficient of determination R^2 can be determined by equation [14] as follows:

$$R^2 = \frac{\mathrm{SS_{Reg}}}{\mathrm{SS_{Corr}}} = \frac{\sum_{t=1}^{n}(\hat{Y}_t - \bar{Y})^2}{\sum_{t=1}^{n}(Y_t - \bar{Y}_t)^2}. \tag{15}$$

If the value of coefficient of determination $R^2 = 1$, it means that the percentage of independent variable contribution to the dependent variable Y_t amount 100%. If the value of the coefficient of determination $R^2 = 0$, it means that the independent variable does not contribute to the dependent variable Y_t.

IOP Conf. Series: Materials Science and Engineering **621** (2019) 012015 doi:10.1088/1757-899X/621/1/012015

2.4. Level of accuracy of forecasting

To measure the error rate of the model estimator in predicting a value, it can be measured based on the error generated from the forecast. According to Sukono et al. [14], to measure the level of forecast, error is usually done by means of Mean Square Error (MSE). MSE is the average size of the quadratic value of forecasting errors. To determine the value of MSE can be done using the equation as follows:

$$MSE = \frac{\sum_{t=1}^{n} \varepsilon_t^2}{n} = \frac{\sum_{t=1}^{n} (Y_t - \hat{Y}_t)^2}{n}. \qquad (16)$$

A model is said to be appropriate, if the error resulting from the prediction is very small close to zero or its accuracy rate approaches 100%.

3. Results and Anaysis

In this section, we discussed the result and analysis, which included data analyzed, model parameter estimation, model estimator significance test, deterministic coefficient determination, and measurement of model accuracy in forecasting.

3.1. Data analyzed

The data analyzed in this study is secondary data, i.e. GDP data based on constant prices, population, and CO_2 emissions 1967-2014. The data is obtained from the World Bank's official website (https://data.worldbank.org/). Descriptive statistics are presented with a view to providing an overview of the quantitative data used in the study. The presentation of descriptive statistical data in this study was conducted using Microsoft Excel 2013. The descriptive statistics can be seen in Table 1. Where Y_t is the CO_2 emissions, Z_{1t} is GDP at constant prices, and Z_{2t} is the total population.

Table 1. Descriptive statistics

Statistic	Y_t	Z_{1t}	Z_{2t}
Mean	1.02753743	1838.98859	181239285.4
Median	0.89894000	1750.89300	1.8300000
Maximum	2.55975023	3692.94288	2.5500000
Minimum	0.23191548	656.746824	105907403
Std.Dev	0.57736700	835.768600	446693252

Graphs of CO_2 emissions data, GDP at constant prices, and population numbers are presented in Figure 2.

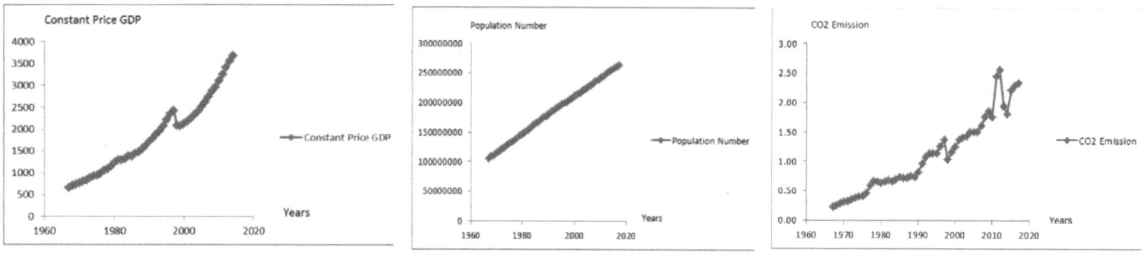

(a) GDP at constant prices (b) Populasi number (c) CO_2 emissions

Figure 2. GDP at current prices, population size, and CO_2 emissions

3.2. Estimation of model parameters

GDP data at constant prices, population and CO_2 emissions obtained from World Bank official website are transformed into natural logarithmic values by reference equation (2). The transformed data is then modeled into the Cobb-Douglas model with reference to (1). Estimation of model parameters was performed using the ant colony optimization (ACO) algorithm developed in section 2.2.

Model parameter estimation process is done by using Matlab R2015a software, the stages are as follows:

- Click start → choose *Matlab* R2015a.
- On the File menu, select *New Script.*
- Then write the objective function along with the CO_2 emissions data as Y_t, GDP at constant prices as Z_{1t}, and the population as Z_{2t} on the sheet script. Then save the *script* by name *obj1_function.m*. The objective function is as given in (4).
- Open *Command Windows* to call the *script, which* has been made in the previous stages.
- To get parameter values based on data already entered into function, run *obj1_function.m* how to click *run* then type *Y =particleswarm(@obj1_function,3)* on *Command Windows* and press *enter*, the execution results appear the estimated parameter values.

Based on the estimation process using ACO obtained values estimator model parameters are respectively $\hat{\theta}_0$= -22.0640901, $\hat{\theta}_1$ = 0.819405999, and $\hat{\theta}_2$ = 0.834930855. Thus, by reference to equation (2), the multiple linear regression model of the estimate can be expressed as follows:

$$Y_t = -22.0640901 + 0.819405999 Z_{1t} + 0.834930855 Z_{2t} + \varepsilon_t \,. \qquad (17)$$

In order to ensure that the model of the estimation result using ACO is to significantly draw the pattern of the analyzed data, it is necessary to test the significance of the model estimator. The testing stages of significance of the model estimator are as follows.

3.3. Testing the significance of model estimators

The model significance significance test performed in this section includes: simultaneous parameter estimator significance test, individual parameter estimator significance test, and assumption of residual normality test.

- Test of simultaneous significance

In the simultaneous test of significant parameters, the hypothesis used is H_0: $\hat{\theta}_0 = \hat{\theta}_1 = \hat{\theta}_2 = 0$, which means the independent variables together do not have a significant effect on the dependent variable. While the alternative hypothesis used is H_1: $\exists\ \hat{\theta}_0 \neq \hat{\theta}_1 \neq \hat{\theta}_2 \neq 0$, which means there are independent variables that influence significantly to the dependent variable.

Statistic value of F_{Stat} is calculated by using equation (11), and the statistic value is obtained $F_{Stat}=$ 853.2750831. Meanwhile, using the level of `significance $\alpha=$ 0.05, and degrees of freedom $df=(k-1;n-k)=$ (2; 45), of the distribution table of F can be obtained the critical value of statistics $F_{(0.05;2;45)}=$ 3.205. Thus, it can clearly be shown that $F_{Stat}>F_{(0.05;2;45)}$, therefore it rejects the hypothesis H_0 and accepts accept the hypothesis H_1. This means that at a significant 5% or 95% confidence level, GDP at constant prices and population totals, or at least a parameter estimator $\hat{\theta}_0=$ -22.0640901, $\hat{\theta}_1$ = 0.819405999, and $\hat{\theta}_2$ = 0.834930855 affect the CO_2 emission-free variable in Indonesia.

- Test of individual significance

In testing the significance of parameter estimators individually, each parameter estimator one by one is tested. Starting with the test for the parameter estimator $\hat{\theta}_0$, where the hypothesis used is H_0 : $\hat{\theta}_0=0$, it means that a constant estimator $\hat{\theta}_0$ has no partial effect on the dependent variable. While the alternative hypothesis is H_1: $\hat{\theta}_0\neq 0$, meaning that a constant estimator $\hat{\theta}_0$ has a partial effect on the dependent variable.

Value of statistics t_{Stat} is determined using equation (12), and the statistical value is obtained $t_{Stat}=$ 3.10370598. Meanwhile, by using a significant level $\alpha=0.05$, and degree of freedom $df=n-k=45$, from the distribution table of t can be obtained critical value statistic $t_{(0.025;45)}=$ 2.01410. Therefore, it can be shown that $|t_{Stat}|>|t_{(0.025;45)}|$, rejects the hypothesis H_0 and therefore accepts the hypothesis H_1. This Means that with a significant level of 5% or a 95% confidence level, the parameter estimator is a constant $\hat{\theta}_0=$ -22.0640901 individually has an effect on the CO2 emission-free variable.

Tests of individual significance also need to be made against the estimator coefficient parameters $\hat{\theta}_1$ and $\hat{\theta}_2$. Using the same way, the test results show that the parameter estimator coefficients $\hat{\theta}_1$ = 0.819405999, and $\hat{\theta}_2$ = 0.834930855, significantly each also has an effect on the CO_2 emission-free variable.

- Test of the assumption of residual normality

Testing assumption of residual normality in this paper, conducted by using statistical test Kolmogorov-Smirnov. In testing the assumption of this residual normality, the hypothesis used is H_0: the residual data follows the normal distribution. While the alternative hypothesis is H_1: the residual data does not follow the normal distribution.

Determination of statistical value of Kolmogorov-Smirnov D_{Stat} can be done by using equation (13), and obtained statistical value $D_{Stat}=0.07004871$. Meanwhile, using the level of significance $\alpha=0.05$ and the number of data $n=48$, from the Kolmogorov-Smirnov table obtained the critical value of statistics $D_{(0.05;48)}=0.198$. Therefore, it can be shown that $D_{Stat}>D_{(0.05;48)}$, thus rejecting the hypothesis H_0 and therefore accepting hypothesis H_1. This means that the residual data follows the normal distribution. From the process of testing the residual normality also obtained value estimator mean $\hat{\mu}_\varepsilon=$ 0.005327, and variance estimator $\hat{\sigma}^2_\varepsilon=$ 0.015567. Thus, it can be shown that $\varepsilon_t\sim$ $N(0.005327, 0.015567)$.

3.4. Determination of the value of the coefficient of determination

In this section, calculating the coefficient of determination can be used to measure the strength of the correlation between independent variables with the dependent variable. Value of coefficient of determination R^2 using equation (15), and the result is $R^2 = 0.980211005$. This means that the large contribution of GDP at constant prices and population to the variation (ups and downs) of CO_2 emissions by 98%, while the remaining 2% is caused by other factors outside the GDP at current prices and the population of Indonesia.

3.5. Measurement of the accuracy of forecasting

Measuring the level of accuracy of forecasting is very important, to measure the accuracy of the model estimator when used for forecasting. It is known that the model obtained from the estimation process using ACO is given as in (17). If it refers to equation (3), then equation (17) means having the model estimator as follows:

$$\hat{Y}_t = -22.0640901 + 0.819405999 Z_{1t} + 0.834930855 Z_{2t}$$

While it is known that $Y_t = \ln G_t$; $\theta_0 = \ln \phi$; $Z_{1t} = \ln X_{1t}$; and $Z_{2t} = \ln X_{2t}$, so this last equation can be expressed as follows:

$$\hat{G}_t = e^{-22.0640901} X_{1t}^{0.819405999} X_{2t}^{0.83493085} . \tag{18}$$

In equation (18), the amount of elasticity $\hat{\theta}_1 + \hat{\theta}_2 = 1.654336854$, show that $\hat{\theta}_1 + \hat{\theta}_2 > 1$. Therefore, equation (18) is a Cobb-Douglas production function that has a characteristic increasing return to scale. This means that there is an increase in the values of the proportion of input units X_{1t} and X_{2t}, will lead to an increase in the proportion of output of the CO_2 emission-free variable doubled from the proportion of the input unit X_{1t} and X_{2t}. Furthermore, equation (18) is used to forecast the data out sample. The objective is to measure the accuracy of model estimators when used in forecasting.

Measurements are performed starting with the determination of the error rate, and then used to calculate the accuracy of the model estimator. Determination of error value can be done by using data in sample. The error value is calculated by using the MSE as given in equation (16), and the value of MSE is obtained 0.01521019 or 1.521019%. Therefore, it can be interpreted that the accuracy of the model estimator of GDP at constant prices and the population on CO_2 emissions, using the ACO algorithm is 0.98478981 or 98.478981%. The graph of actual data and forecasting results as given in Figure 3, as follows:

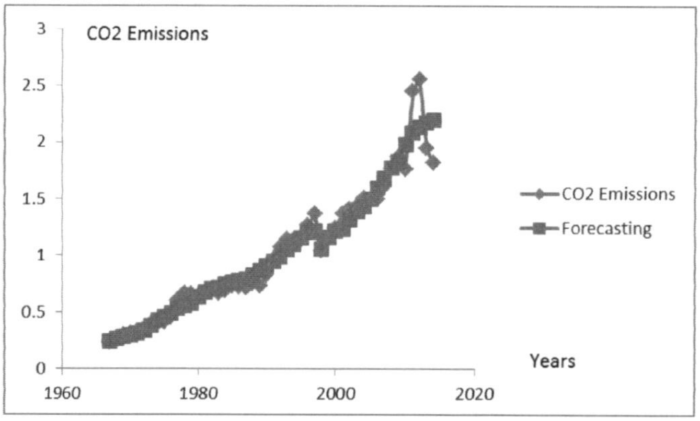

Figure 3. Data of actual and forecasting in sample

Furthermore, if we estimate the approximate values X_{1t} and X_{2t} as given in Table 2, then forecasting out sample values of CO_2 emissions by 2015, 2016, 2017, and 2018 can be done. Incorporates the values of each variable that increase from the previous year, and forecasting is done using equation (18) the result as given in Table 2, column $\hat{G}_t$.

Table 2. Forecasting out sample value of emissions CO_2

Years (t)	X_{1t}	X_{2t}	$\hat{G}_t$
2015	3825.77917	258229969	2.381406066
2016	3958.615465	261328822	2.473462628
2017	4091.45176	264427675	2.566405774
2018	4224.288055	267526528	2.660236089

From Table 2, it appears that there is an increase of input unit in the form of GDP based on constant prices and population, causing an increase in CO_2 emissions that multiply in Indonesia.

4. Conclusion

In this paper, we has been done in estimating the model of the impact of GDP growth at constant prices and population on CO_2 emissions, using the Cobb-Douglas model of the parameters estimated using ant colony optimization (ACO). Based on the analysis, it can be concluded that the impact of the growth of GDP at constant prices and population on CO_2 emissions is significantly following the Cobb-Douglas production function with elasticity estimator $\hat{\theta}_1$ = 0.819405999, and $\hat{\theta}_2$ = 0.834930855. Because $\hat{\theta}_1 + \hat{\theta}_2$ = 1.654336854, show that $\hat{\theta}_1 + \hat{\theta}_2 > 1$, the obtained production function estimator has the characteristics of increasing return to scale. This indicates that the proportion of output additions doubled from the proportion of input additions. Furthermore, forecasting based on data in sample, error rate measured using MSE obtained value of 0.01521019 or 1.521019%. This means that the estimator accuracy rate of the model estimator of the impact of GDP at constant prices and the population on CO_2 emissions, using the PSO algorithm is 0.98478981 or 98.478981%. This suggests that the Cobb-Douglas production function estimator obtained is relatively accurate to be used for increasing CO_2 emissions, as a result the growth of GDP at constant prices and population. Taking into account such conditions of high CO_2 emissions, such modeling is expected to be taken into consideration for the authorities in policy making in controlling the increase of CO_2 emissions in Indonesia.

5. Acknowledgements

Acknowledgments are conveyed to the Rector, Director of Directorate of Research, Community Involvement and Innovation, and the Dean of Faculty of Mathematics and Natural Sciences, Universitas Padjadjaran, with whomthe Internal Grant Program of Universitas Padjadjaran was made possible to fund this research.The grant is a means of enhancing research and publication activities for researchers at Universitas Padjadjaran.

References

IOP Conf. Series: Materials Science and Engineering **621** (2019) 012015 doi:10.1088/1757-899X/621/1/012015

[1] Alam J 2014 On the Relationship Between Economic Growth and CO_2 Emissions: The Bangladesh Experience. *IOSR Journal of Economics and Finance (IOSR-JEF)*. Volume 5, Issue 6, pp-36-41. E-ISSN: 2321-5933.

[2] Aye G C and Edoja P E 2017 Effect of Economic Growth on CO2 Emission in Developing Countries: Evidence from a Dynamic Panel Threshold Model. *Cogent Economics & Finance*(2017), 5: 1379239, pp. 1-22. https://doi.org/10.1080/23322039.2017.1379239.

[3] Conrad E and Cassar L F 2014 Decoupling Economic Growth and Environmental Degradation:Reviewing Progress to Date in the Small Island State of Malta. *Sustainability* 2014, *6*, pp. 6729-6750; doi:10.3390/su6106729.

[4] Duan P and Yong A I 2016 Research on an Improved Ant Colony Optimization Algorithm and its Application. *International Journal of Hybrid Information Technology*. Vol.9, No.4 (2016), pp. 223-234. http://dx.doi.org/10.14257/ijhit.2016.9.4.20.

[5] Felipe J and Adams F G 2005 The Estimation of the Cobb-Douglas Function: A Retrospective View. *Eastern Economic Journal*, Vol. 31, No. 3, Summer 2005, pp. 427-445.

[6] Lacheheb M, Rahim A S A and Sirag A 2015 Economic Growth and Carbon Dioxide emissions: Investigating the Environmental Kuznets Curve Hypothesis in Algeria. *International Journal of Energy Economics and Policy*, 2015, 5(4), 1125-1132.

[7] Mehrizi M A, Atashi S M and Elahi M 2012 The effect of variables population growth, urbanizationand economic growth on CO2 Emissions in Iran. *African Journal of Business Management* Vol.6 (28), pp. 8414-8419, 18 July, 2012, pp. 8414-8418. Available online at http://www.academicjournals.org/AJBM. DOI: 10.5897/AJBM11.2020. ISSN 1993-8233 ©2012 Academic Journals.

[8] Mitic P, Ivanovic O M and Zdravkovic A 2017 A Cointegration Analysis of Real GDP and CO2 Emissions in Transitional Countries. *Sustainability*2017, 9, 568, pp. 1-18. doi:10.3390/su9040568.

[9] Obradovic S and Lojanica N 2017 Energy Use, CO2 Emissions and Economic Growth-Causality on a Sample of SEE Sountries. *Economic REsEaRch-Ekonomska istRaživanja*, 2017, voL. 30, no. 1, 511–526. http://dx.doi.org/10.1080/1331677X.2017.1305785.

[10] Omri A 2013 CO2 Emissions, Energy Consumption and Economic Growth Nexus in MENA Countries: Evidence from Simultaneous Equations Models. *Energy Economics* 40 (2013) 657–664. http://dx.doi.org/10.1016/j.eneco.2013.09.003.

[11] Salami, Al N M A 2009 Ant Colony Optimization Algorithm. *UbiCC Journal*, Volume 4, Number 3, August 2009, pp. 823-826.

[12] Samsami R 2013 Application of Ant Colony Optimization (ACO) to forecast CO2 Emission in Iran. *Bulletin of Environment, Pharmacology and Life Sciences. Bull. Env. Pharmacol. Life Sci.*, Vol 2 (6) May 2013: 95- 99. ©2013 Academy for Environment and Life Sciences, India. Online ISSN 2277-1808. Journal's URL:http://www.bepls.com.

[13] Sasana H and Putri AE 2018 The Increase of Energy Consumption and Carbon Dioxide (CO2) Emission in Indonesia. *E3S Web of Conferences*31, 01008 (2018), ICENIS 2017, pp. 1-5. https://doi.org/10.1051/e3sconf/20183101008.

[14] Sukono, Hidayat Y, Suhartono Sutijo B BonA.T.B., and Supian, S. 2016. Indonesian Financial Data Modeling and Forecasting By Using Econometrics Time Series and Neural Network. *Global Journal of Pure and Applied Mathematics*. ISSN 0973-1768 Volume 12, Number 4 (2016), pp. 3745-3757. © Research India Publications. http://www.ripublication.com/gjpam.htm.

[15] Yang L, Yuan S and Sun L 2011 The Relationships between Economic Growth and Environmental Pollution Based on Time Series Data:An Empirical Study of Zhejiang Province. *Journal of Cambridge Studies,* Volume 7, No. 1, pp. 33-42.

[16] Yue and Wang X 2015 An Improved Ant Colony Optimization Algorithm for Solving TSP. International *Journal of Multimedia and Ubiquitous Engineering* . Vol.10, No.12 (2015), pp.153-164. http://dx.doi.org/10.14257/ijmue.2015.10.12.16.

[17] Zarzoso I. M, Morancho A B and Lage R M 2006 The Impact of Population on CO2 Emissions: Evidence from European Countries. *Paper.* NOTA DI LAVORO 98.2006. web site: www.feem.it.

[18] Zhao Y and Wang S 2015 The Relationship between Urbanization, Economic Growth and Energy onsumption in China: An Econometric Perspective Analysis. *Sustainability* 2015, *7*, pp. 5609-5627; doi:10.3390/su7055609.

[19] Zhu Q and Peng X 2012 The Impacts of Population Change on Carbon Emissions in China During 1978–2008. *Environmental Impact Assessment Review* 36 (2012) 1–8. journal homepage: www.elsevier.com/locate/eiar.

Comparative analysis of inline quality control system process capability with offline lab QC against product specification by MSPC method

Yuyun Hidayat[1], Sri Winarni[1], Titi Purwandari[1], Anita Sukmawati[1], Sukono[2]

[1]Department of Statistics, Faculty of Mathematics and Natural Sciences, Universitas Padjadjaran, Indonesia (email: yuyun.hidayat@unpad.ac.id, sri.winarni@unpad.ac.id, titipurwandari@yahoo.com, anita.241195.as@gmail.com)

[2]Department of Mathematics, Faculty of Mathematics and Natural Sciences, Universitas Padjadjaran, Indonesia (email: sukono@unpad.ac.id)

*corresponding author: sukono@unpad.ac.id

Abstract. The similarity of values shown by the two methods of reading paper specifications, namely Quality Control System (QCS) and QC Lab (controlling offline in Quality Control lab) to the attention of PT Bukit Muria Jaya (BMJ). The QCS is an online control process while QC Lab is offline control process. The company wants to change the offline control process online, but there is no analysis to show whether the output given by both methods is the same. The study involved Grade A paper measured at once by five paper specification parameters, i.e. paper weight, porosity, CaCo3 content, opacity, and paper strength. However, after normality testing was found, the data is not multivariate normal distribution. Therefore, this study aims to answer the problem with a multivariate free distribution procedure. The study was conducted in two stages, firstly by looking at the median equality of both outputs of the paper specification reading method, and the second is by looking at the similarity of characteristic capability. A preliminary study was conducted to see whether the two methods of reading the Grade A paper specifications had the same median. This test is done by Kruskal Wallis Multivariate method. The results of the analysis received a null hypothesis which means the median of the two outputs generated QCS and QC Lab for each Grade A paper are equal. Furthermore, based on a capability comparison analysis where the value of this capability is derived from the sample without observation out of control, out of control process control is assisted in the range of T2 Hotelling Control Free Distribution. It can be concluded that Grade A paper capability of QCS does not show the same result as QC Lab, i.e. Porosity variable, where Porosity is capable in QC Lab but not QCS, so manual process control for Grade A paper cannot yet be replaced online.

Keywords: Grade A paper, normality testing, Kruskal Wallis multivariate method, T^2 Hotelling

1. Introduction

Paper in this company using Popereel Machine (PM), currently there are three PM PM1, PM2, and PM3. In this case the researcher will take data from PM3. On the PM is installed software paper controller that is being produced. Control with this software is a control that is done online. One of the installed software is Quality Control System (QCS) which focuses on reading the paper specification being processed. Control is also done manually in QC Lab, which is a place to analyze product specifications through manual sampling using a calibrated tool. The research data was taken from PM3 in November 2016 with Grade A. In that month, there was a gap between QCS and QC Lab, the gap indicated that the value of gap exceeded the company's tolerance limit. The largest gap is in the porosity variable of 3.88 points [14; 15].

Taken the average of each variable as shown in Table 1, it turns out that the average value gap given by both paper specification methods indicates that there are variables that tend to exceed the tolerance limit i.e. porosity variable of 2.0830 points.

Table 1. Average Gap Value of each Variable in Each Popereel Machine (PM)

PM3 (*Grade* A)				
Contents $CaCo_3$	Paper Weight	Paper Strength	Opacity	Porosity
0.253241	0.427701	1.983982	0.716505	2.083039

Is this company's tolerance limit a good benchmark for concluding differences in data to be of interest to the researcher. Since these variables come from the same observation of reels or rolls of paper, the variables must be tested by multivariate analysis whether QCS significantly gives different output to the QC Lab for the gap. Currently there is no research that conducted the test on PT Bukit Muria Jaya (BMJ). Yet this gap is a concern for companies to use QCS data as data used to describe the company's production process [14; 15].

Leoni et al. [9], investigates the effect of correlation and autocorrelation in the Hotelling T^2 graph when there are two quality characteristics of X and Y, whose autocorrelation structure and correlation are represented by the VAR model (1). Their research is done by using Mahalanobis distance and geometric approach with ellipse. The results of his research indicate that the presence of autocorrelation reduces the performance of T^2 charts, and limits the ability of diagrams to provide specific cause information applicable in the process. A higher or lower degree of autocorrelation is the performance of the graph in detecting the cause of the change in value on the average vector. Several examples have been presented to illustrate the adverse effects of T^2 charts when there is autocorrelation in the process. Ramzan et al. [11], conducting quality control studies and activities with a special focus on offline inspection. In his research various models and methodologies were developed for offline inspection under different manufacturing and inspection conditions. Their studies are classified into six groups based on their research objectives, developed models, adopted methodologies, and research results. In their research also provides a brief overview of offline inspection to propose future research opportunities and emerging trends. The proposed research may assist in the development of new models or modify existing models to improve offline inspection performance. Similar research on the capability of the process of quality control system has been done by Ferrer [5], Hadad & Alsmadi [6], Kormaz et al. [7], Lee [8], Hidayat et al. [14; 15] and Zhan et al. [16].

Therefore, based on the above description, the researcher is interested to analyze whether there is significant difference of output from QCS when compared with QC Lab. While the purpose of this research is to replace the control of offline production into online that is by using QCS. So from this research for PT Bukit Muria Jaya is can be used as material evaluation of QCS (Quality Control System) that is being applied company. If the QCS and QCS Labs show the average similarity and show the similarity of variations based on the customers' specifications then QCS is feasible to be used as a trusted process control.

2. Methodology

Based on the background of research in the previous chapter, it is known that mathematically there is a difference in the output given between QCS software and QC Lab analyst. This difference does not indicate a good or bad thing for the company when viewed mathematically. Therefore this difference will be tested statistically in order to provide useful information for the company. The supporting theories to conduct an analysis of the problem will be described as follows.

2.1. Description of research data

Data obtained from Grade A paper produced in PM 3 in November 2016, sampling carried out every 30,000 meters of paper length. The sample is taken from the paper in production, i.e. at the last 4 meters before the length to 30,000 meters. The paper specifications of this sample are then calculated by two methods, namely QCS and QC Lab. In each paper specification variable taken 3 points of observations. The first sample is computed into the first subgroup.

Involved five of the same paper characteristic variables in the two paper reading methods as follows: X_1= paper weight/grammature (in gr/m^2); X_2= paper strength/size press moisture (in %); X_3 = CaCo3/ash titration content (in %); X_4 = opacity/opacity (in %); and X_5 = porosity/permeability (in coresta).

2.2. Assumption testing

Assumption tests performed in this section include: correlation testing with Bartlett sphericity, and assaying multivariate normality assumptions.

- Correlation test with Bartlett sphericity

This correlation test aims to show whether there is a significant correlation between the five variables involved in both paper readout methods. According to Lim et al. [10]. This test can be performed using Bartlett sphericity testing with the following steps.

Develop a statistical hypothesis that is H_0: $\boldsymbol{R} = \boldsymbol{I}$ and H_1: $\boldsymbol{R} \neq \boldsymbol{I}$; then calculate chi-square test statistic according to equation as follows:

$$\chi^2_{hitung} = -m - 1 - \frac{2p+5}{6} \ln |R| \tag{1}$$

where m is the number of observations, and $\boldsymbol{R}$ is the correlation matrix of $\mathbf{x}$ with as many variables p.

The testing criterion is rejecting the null hypothesis if the value $\chi^2_{hitung} \geq \chi^2_{\alpha;1/2p(p-1)}$ or p-value $\leq \alpha$, meaning that the correlation matrix obtained is not the identity matrix so that the characteristic variable of paper specification correlates each other. Therefore the use of multivariate methods is feasible to use.

- Testing multivariate normality assumptions

Testing the assumption of normality is assisted by using package "MVN" in application R with hypothesis as follows:

$H_{01}: b_{1,p} = 0$ (data is normally distributed)

$H_{11}: b_{1,p} \neq 0$ (data is not normally distributed)

$H_{02}: b_{2,p} = 0$ (data is normally distributed)

$H_{12}: b_{1,p} \neq 0$ (data is not normally distributed)

Mardia's MVN test is an assay method of normal multivariate distribution based on the skewness extension $b_{1,p}$ and kurtosis $b_{2,p}$ with skewness test statistics z_1 and kurtosis test statistics z_2 as follows:

$$z_1 = \frac{(p+1)(m+1)(m+3)}{6\{(m+1)(p+1)-6\}} b_{1,p} \tag{2}$$

$$z_2 = \frac{b_{2,p}-p(p+2)}{\sqrt{8p(p+2)/n}} \tag{3}$$

where

$$b_{1,p} = \frac{1}{m^2}\sum_{i=1}^{m}\sum_{j=1}^{m} g_{ij}^3 \tag{4}$$

$$b_{2,p} = \frac{1}{m}\sum_{i=1}^{m} g_{ii}^2 \tag{5}$$

where $g_{ij} = (x_i - \bar{x})'S^{-1}(x_j - \bar{x})$ distance of mahalanobis square of *i*-th variable and variable to-j; $g_{ii} = (x_i - \bar{x})'S^{-1}(x_i - \bar{x})$; p = the number of related variables; m = the number of multivariate observations; and S = covariance Matrix [4; 12].

Testing criterion is rejected H_{01} when $z_1 \geq \chi^2_{\alpha,p(p+1)(p+2)/6}$. While H_{02} show kurtosis or in other words show whether the distribution of data is too uphill or not with the criterion test rejected H_{02} if $z_2 \geq z_{\alpha/2}$.

2.3. *Multivariate average equality test*

The multivariate average equality tests performed in this section include: one-way manova and expansion of Kruskal Wallis test for multivariate data.

- One-way manova

This admission can be done if the assumption of homogeneity of variance and multivariate normal distribution is met, where the homogeneity test of variance can be done by M-box test, but if both assumptions are not met then non-parametric statistics can be performed. The test statistic used is Wilks Lamda according to the equation:

$$\Lambda^* = \frac{\left|\sum_{l=1}^{g}\sum_{j=1}^{n_l}(x_{lj}-\bar{x}_l)(x_{lj}-\bar{x}_l)'\right|}{\left|\sum_{l=1}^{g}\sum_{j=1}^{n_l}(x_{lj}-\bar{x})(x_{lj}-\bar{x})'\right|} \tag{6}$$

where $\boldsymbol{x}_{lj}$ = observation method of reading the paper specifications to-l variable characteristics of paper to-j; $\bar{\boldsymbol{x}}_l$ = average vector paper reading method for paper specification to-l; and $\bar{\boldsymbol{x}}$ = the average vector of both paper reading method specifications [2].

Test criteria reject the null hypothesis if $\left(n - 1 - \frac{p+g}{2}\right) ln(\Lambda^*) \geq \chi^2{}_{p(g-1)}$

- Expansion of Kruskal Wallis test for multivariate data

This test is performed if the assumption of normal distribution cannot be met. This test is conducted to find whether there is a difference of distribution between QCS and QC Lab. The steps to perform this test are as follows [3].

Develop a statistical hypothesis H_o : med $(\underline{X})_a$ = med $(\underline{X})_b$ and H_1 : med $(\underline{X})_a \neq$ med $(\underline{X})_b$; $a \neq$ b. Where a = method of reading lab paper QC specifications, and b = QCS paper specification reading method. Calculate the value of the test statistic KW with the following formula:

$$KW = \sum_{k=1}^{K} n_k(\bar{R}^{(k)})'\hat{\Sigma}_n^{-1}\bar{R}^{(k)} \tag{7}$$

$$\bar{R}^{(k)} = \frac{1}{n_k}\sum_{i\epsilon N_k} R_k(x^{(i)}) \tag{8}$$

$$R_k(x^{(i)}) = \frac{1}{n_k}\sum_{j=1}^{n_k} sign(x^{(i)} - x^{(j)})$$

$$sign(x^{(i)} - x^{(j)})\begin{cases}-1\ jika\ (x^{(i)} - x^{(j)}) < 0\\ 0\ \ jika\ (x^{(i)} - x^{(j)}) = 0\\ 1\ \ jika\ (x^{(i)} - x^{(j)}) > 0\end{cases} \tag{9}$$

$$\hat{\Sigma} = \frac{1}{n-K}\sum\sum_{k=1,i\epsilon N_k}^{K} R_k(x^{(i)})R_k(x^{(i)})' \tag{10}$$

where N_k many observations on the reading method to-*k*; n_k many observations of each variable on the reading method to-*k*; k paper specification reading method (QC Lab, QCS); and $R_k(x^{(i)})$ vector size scale ranking method of reading paper specification to-*k*.
Test criteria for the calculation, reject the null hypothesis if $KW \geq \chi^2_{\alpha,k-1}$. Level of significance α used is 5% or 0.05.

2.4. Multivariate control chart T^2 Hotelling free distribution

From the checking assumption of multivariate normality, if the normal multivariate assumption is not met, then the control chart formed is a control chart T^2 Hotelling free distribution. This diagram can be formed according to the steps below [1; 14].

$$\mathrm{T}_i^2 = n(\bar{x} - \bar{\bar{x}})'S^{-1}(\bar{x} - \bar{\bar{x}}) \quad (11)$$

$$\mathrm{UCL} = \overline{T^2} + ks_{T^2} \quad (12)$$

Create a control chart by creating a plot between values T_i^2 (i = 1,2,3,…,n) with UCL.

$$\bar{x}_{jk} = \frac{1}{n}\sum_{i=1}^{n} x_{ijk} \begin{cases} j = 1,2,\ldots,p \\ k = 1,2,\ldots,m \end{cases} \quad (13)$$

If the value T_i^2 is outside the control limit (UCL) then the observations are declared uncontrollable
.

2.5. Process capability analysis

Good capability analysis is used for normally distributed data but when data not normally distributed we use the following equation:

$$C_{pc} = \frac{USL-LSL}{6\sqrt{\frac{\pi}{2}}E|X-T|} \quad (14)$$

where $E|\mathrm{X} - T|$ obtained through $E|\mathrm{X} - T| = \frac{1}{n}\sum_{i=1}^{n}|x_i - T|$; x_i = process row to-*i*; and T = target company [13; 15].

The capability value of Equation (14) can be used to predict the process performance but can only be used if the process is declared in control. It is necessary to obtain the in control process by making a revision control chart without any uncontrolled observation (out of control).

Capability calculations are performed for each characteristic variable of paper in QCS and QC Lab. Then the capability value obtained is compared through hypothesis testing, whether the capability value given from QCS is the same as compared to QC Lab.

2.6. Capability comparison

Steps in performing capability comparison are as follows: Compiled the statistical hypothesis of capability comparison with comparator 1.33, and Finding critical value C. The C critical value obtained with the following formula [15]:

$$C = \frac{\mathrm{C}}{\mathrm{Cp}\ low} \times \mathrm{Cp}\ low \quad (15)$$

where the value $\frac{\mathrm{C}}{\mathrm{Cp}\ low}$ obtained from table of sample size and critical value of determination based on the value of A with α=5%. A value is obtained by using the formula:

$$A = \frac{\mathrm{Cp}\ high}{\mathrm{Cp}\ low} \quad (16)$$

where Cp $high$ = process capability to be received with opportunities 1-α; Cp low = process capability to be rejected with opportunities 1-β; Cp $high$ = $\max(1.33, \mathrm{Cpc(QCS)})$ or $\max(1.33, \mathrm{Cpc(QC\ Lab)})$; and Cp low = $\min(1.33, \mathrm{Cpc(QCS)})$ or $\min(1.33, \mathrm{Cpc(QC\ Lab)})$.

The test criterion is $\hat{C}_p$ compared to Critical C, with the criterion of the starting test H_0 when $\hat{C}_p >$ critical value C $(\alpha = \beta = 0.05)$. Where $\hat{C}_p$ is the process capability shown by QCS and QC Lab.

$$C = \frac{\mathrm{C}}{\mathrm{Cp}\ low} \times \mathrm{Cp}\ low \tag{17}$$

where the value $\frac{\mathrm{C}}{\mathrm{Cp}\ low}$ obtained from table of sample size and critical value of determination based on the value of A with α=5%. The value of A is obtained by using the formula:

$$A = \frac{\mathrm{Cp}\ high}{\mathrm{Cp}\ low} \tag{18}$$

$\hat{C}_p$ compared to Critical value of C, with the criterion of the starting test H_0 when $\hat{C}_p >$ critical value C $(\alpha = \beta = 0.05)$. Where $\hat{C}_p$ is the process capability shown by QCS and QC Lab [14; 15].

3. Results and discussion

Based on the analysis steps and problems that have been described in the previous chapter, in this chapter will be discussed about the analysis of the problems. This analysis is assisted by Microsoft Office Excel software and R software.

3.1. Correlation Testing

Before the analysis begins, the first test is to test the correlation assumption between the Grade A paper measurement variables, if there is a correlation between the corresponding variants then the multivariate test is feasible to use. There are 5 variables that will be tested each from QCS reading method and QC Lab reading method that is variable of paper weight, axis, CaCo3 content, transparency, and paper strength. It is said that these variables are correlated if they do not form an identity matrix. So the null hypothesis is $H_0\!: \boldsymbol{R} = \boldsymbol{I}$ (the correlation matrix is the same as the identity matrix, meaning the paper quality specification variable has no correlation). While the one hypothesis is $H_1\!: \boldsymbol{R} \neq \boldsymbol{I}$ (the correlation matrix is not the same as the identity matrix, meaning the paper quality specification variables correlate each other). Thus, using the formula (1) can be calculated as follows:

$$\chi^2_{QCLab} = -68 - 1 - \frac{2(5)+5}{6}\ \mathbf{ln}\ |0.642806| = \mathbf{28.50332}$$

$$\chi^2_{QCS} = -68 - 1 - \frac{2(5)+5}{6}\ \mathbf{ln}\ |0.457049| = \mathbf{50.50116}$$

Based on the results of these calculations can be concluded that both have a value greater than $\chi^2_{0.05;(\frac{1}{2})\times(5(5-1))} = 18.30704$, then the null hypothesis is rejected which means the correlation matrix is not the same as the identity matrix or the paper quality specification variables correlate each other, then a multivariate analysis is feasible to use.

3.2. Testing of multivariate normality assumption

Testing the assumption of normality is necessary, because it will determine the method of analysis used in accordance with previously described. At this stage will be examined the form of quality data distribution paper produced in PT BMJ whether the distribution is normal or not. The data used is the average data of Grade A paper samples per subgroup. There are as many as 68 subgroups of 5 variables which will be seen from two paper specification reading methods (QCLab and QCS). Testing the assumption of multitask normality is aided by software R using the equations (2) to (5) each used to test skewness and kurtosis.

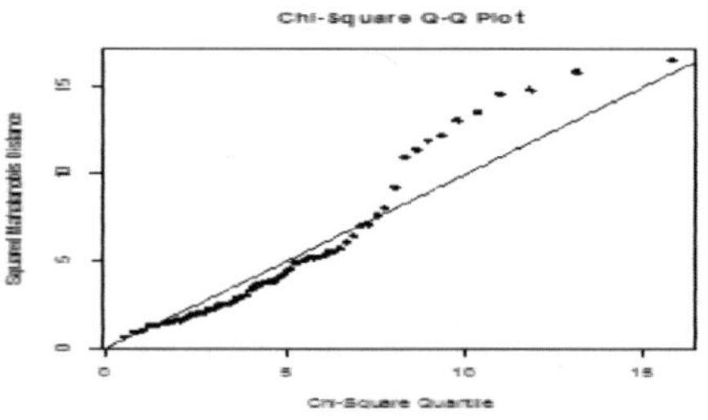

Figure 1. Q-Q Plot QC Lab

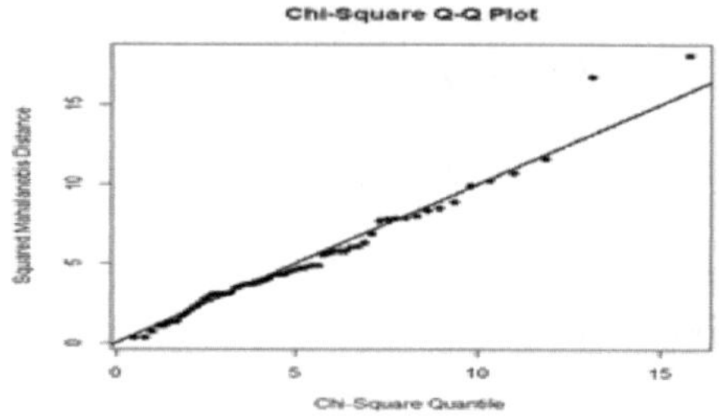

Figure 2. Q-Q Plot QCS

From the subgroups of 68 and two different populations, namely QCS and QC Labs with each corresponding 5 interrelated variables, the following outputs are obtained in Table 2.

Table 2. Statistical Calculation of the Mardia Test

Description	QC Lab	QCS
z_1	69.07942	91.26461
z_2	0.2436779	3.120264
Conclusion	Non-Normal Multivariat	Non-Normal Multivariat

Based on visual testing and using test statistic it is known that the data is not multinormal distributed. Based on the above results, it can be concluded that Grade A paper quality data obtained by QCS paper specification method and QC Lab paper specification reading method are not multivariate normal distribution, this is because the production of PT BMJ is unstable from subgroup to subgroup. Where, subgroups are sampled data taken every production reaches the length of paper along the 30,000 meters.

3.3. Kruskal Wallis multivariate nonparametric test

Because the assumption of normality is not met, parametric procedures will not be done, because it will make it difficult to draw conclusions. Therefore, the test is followed by a nonparametric procedure where the data is converted to ordinal scale by making it in rank. The method to be used is Kruskal Wallis for multivariate data with hypothesis as follows: hyphotesis H_0 : med $(\underline{X})_{QCS}$ = med $(\underline{X})_{QC\ LAB}$, meaning that the median of both paper-reading methods does not give the difference in paper specification values. While the hyphotesis H_1 : med$(\underline{X})_{QCS} \neq$ med $(\underline{X})_{QC\ LAB}$, meaning that the median of both paper-reading methods gives different paper value specifications.

From equation (7) obtained value $KW = 8.7064 \times 10^{-29} < \chi^2_{0.05,5-1}$ =3.841458821 which means by the same sample size, the output produced by both methods of reading paper specifications on Grade A is the same based on the median.

3.4. Making the T^2 Hotelling control charts free distribution

Making this control diagram is intended to get a controlled process, where out-of-control observation will be eliminated, so that later process capability can be calculated through controlled observation. The calculation of the value of equation (11) uses the mean of equation (12), and the average matrix of covariance variance uses excel. Then the calculation of the statistical value of T^2 Hotelling on equation (11) and the control limit of equation (13) is calculated using the R software.

IOP Conf. Series: Materials Science and Engineering **621** (2019) 012016 doi:10.1088/1757-899X/621/1/012016

Then the statistical value of T^2 Hotelling is compared with the control limit in equation (13) with p=5 then the value $\alpha' = 1 - (1 - 0.05)^5 = 0.2262$, thus obtained value $k = \sqrt{\frac{1}{0.05}} = 2.102497$, which means the control limit is at interval 2.102497 standard snail T_i^2 of the average T_i^2. Constants k this will be used for UCL formation in both paper specification reading methods. Value T_i^2 and UCL are then formed into a control chart for each QC Lab and QCS and the results are as follows:

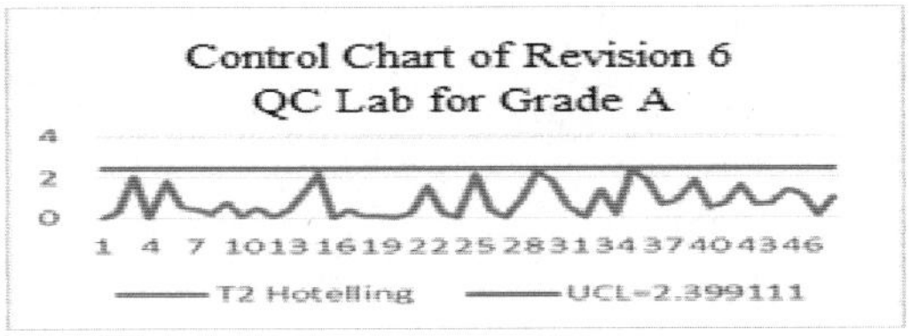

Figure 3. Control chart of revision 6

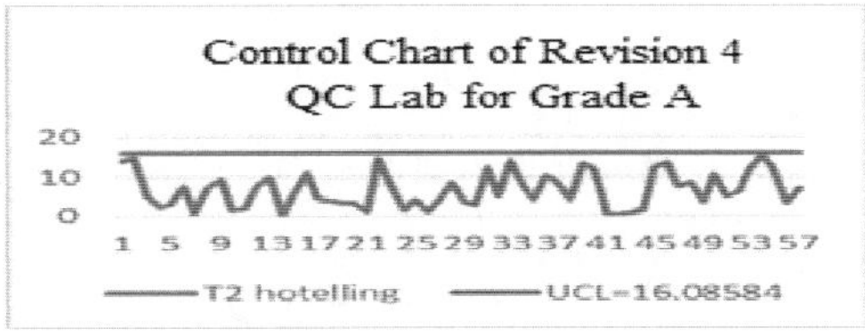

Figure 4. Control chart of revision 4

Uncontrolled uncontrolled data will be used for capability calculation. Capability calculation results are shown in Table 3 and Table 4.

Table 3. Calculation of Variable Capability Values Characteristics of QC Lab Paper

	Variable	*Target Value*	*Upper*	*Lower*	Difference	E|X-Target|	C_{pc}
QC LAB	Contents CaCo3	18	18.5	17.5	1	0.185347	0.717324
	Paper Weight	25	26.5	23.5	3	0.518056	0.769921
	Paper Strength	5	6.5	3.5	3	1.157986	0.344445
	Opacity	74	100	70	30	1.004167	3.97207
	Porosity	20	23.2	16.8	6.4	0.585417	1.453504

Table 4. Calculation of Variable Capability Values Characteristics of QCS Paper

	Variable	*Target Value*	*Upper*	*Lower*	Difference	E|X-Target|	C_{pc}
QCS	Contents CaCo3	18	18.5	17.5	1	1.66549	0.079829
	Paper Weight	25	26.5	23.5	3	0.74119	0.538135
	Paper Strength	5	6.5	3.5	3	1.32088	0.301966
	Opacity	74	100	70	30	0.44264	9.010913
	Porosity	20	23.2	16.8	6.4	1.45606	0.584387

3.5. Comparative analysis of process capability

This process capability calculation is performed for each characteristic variable of Grade A paper produced by PT BMJ for both QCS data and Lab QC data. The calculation data is a calculation data with a controlled process. The calculation of this capability value will be presented in table form as follows.

According to Ferrer [5], a process can be said to be capable if it exceeds the value of 1.33. To see if the process of each variable is capable in both methods, then the hypothesis is formulated as follows: hyphotesis H_{0i} : $\hat{C}_i(a) = 1.33$, paper process capability Grade A the *i*-th variable in the reading method a is not capable. While the hyphotesis H_{1i} : $\hat{C}_i(b) > 1.33$, paper process capability Grade A the *i*tch variable on a capable reading method. Where, a=QC Lab, QCS.

Table 5. Results of Process Capability Testing between QC Lab and Value 1.33

	Variable	C_p High	C_p Low	A= $C_{p\,High}$ / $C_{p\,Low}$	d	Distance A x d	C / $C_{p\,Low}$	$\hat{C}_p$ *(QC Lab)*	Critical Value C	Descrip-tion
QC Lab	Contents CaCo3	1.33	0.72	1.85	0.21	0.026	1.286	0.717324	0.92	No
	Paper Weight	1.33	0.77	1.73	-	-	1.210	0.769921	0.93	No
	Paper Strength	1.33	0.34	3.86	0.21	0.332	1.702	0.344445	0.59	No
	Opacity	3.97	1.33	2.99	0.21	0.151	1.521	3.97207	2.02	*Capable*
	Porosity	1.45	1.33	1.09	0.50	0.084	1.026	1.453504	1.37	*Capable*

Table 6. Results of Process Capability Testing between QCS and Value 1.33

	Variable	C_p High	$C_{p\,Low}$	A= $C_{p\,High}$ / $C_{p\,Low}$	d	Distance A x d	C / $C_{p\,Low}$	$\hat{C}_p$ *(QCS)*	Critical Value C	Descrip-tion
QCS	Contents CaCo3	1.33	0.08	16.66	0.21	2.989	4.36	0.08	0.35	No
	Paper Weight	1.33	0.54	2.47	0.21	0.044	1.41	0.54	0.76	No
	Paper Strength	1.33	0.30	4.40	0.21	0.445	1.82	0.30	0.55	No
	Opacity	9.01	1.33	6.78	0.21	0.937	2.31	9.01	3.07	*Capable*
	Porosity	1.33	0.58	2.28	0.21	0.003	1.37	0.58	0.80	No

3.6. Discussion

The critical value of C is obtained by interpolating the value of the table C/$C_{p\,Low}$ and the value of table A, then the critical value of C obtained compared with the value $\hat{C}_p$ (QCS) and $\hat{C}_p$ (QC Lab), where the value $\hat{C}_p$ (QCS) and $\hat{C}_p$ (QC Lab) each representing the value of process capability provided by QCS and QC Lab. The null hypothesis is rejected if $\hat{C}_p >$ critical value $C(\alpha = \beta = 0.05)$.

Table 5, shows the result of testing the process capability between the QC Lab and the value of 1.33 on each variable. Apparently on the paper Grade A Porosity and Opacity variables are said to be capable, because it meets the value of 1.33. However, paper weight, paper strength and opacity on Grade A paper have not fulfilled the value, meaning that the production process running on PT BMJ has not been able to produce Grade A paper products that are in accordance with the specification or production process has not yet had the ability to produce a Grade A paper product accurate.

Table 6, shows the result of testing the process capability between QCS and the value of 1.33 on each variable. It turns out that only the Opacity variable can be said capable, this happens because the reading of Porosity of Grade A paper on QCS is still not accurate so that the resulting product is said not to meet the specifications even though the Grade A paper has met the customers' specifications. This is a dangerous thing for PT BMJ because it will be very detrimental to the company where later Grade A paper that is ready to sell will be declared as reject goods.

Thus, it can be said that QCS has not been capable in reading the specifications of Grade A paper products. PT BMJ will incur losses when using QCS as a paper reading method believed in this condition. Manual process control cannot yet be replaced with online process control.

4. Conclusion

Based on the Kruskal Wallis test, the results of Grade A paper quality measurements on QC Lab and QCS have no differences in median. Simultaneously, they have the same median QCS and QC Lab are from the same population, meaning that different paper-reading methods do not give a difference to the quality measurements of Grade A paper produced. But testing using Kruskal Wallis only represents the size of the central symptoms. Therefore, an in-depth analysis of the testing of variations based on the specification is done by testing the hypothesis of capability comparison. The results of the capability comparison test for Grade A paper show that QCS still cannot replace the QC Lab, because: (i) QCS has a difference with QC Lab in representing the capability value that can be seen from CaCo3 Content variable. In QC Lab, Porosity is significantly expressed capable with capability of 1.453504, whereas in QCS Porosity variable is not capable, where the value of capability given only 0.58; and (ii) QC Labs are more capable than QCS. This can be seen from the amount of capability shown by both paper reading method specifications. The value of capability shown by QC Lab is larger when compared to QCS.

5. Acknowledgements

Acknowledgments are conveyed to the Rector, Director of Directorate of Research, Community Involvement and Innovation, and the Dean of Faculty of Mathematics and Natural Sciences, Universitas Padjadjaran, with whomthe Internal Grant Program of Universitas Padjadjaran was made possible to fund this research.The grant is a means of enhancing research and publication activities for researchers at Universitas Padjadjaran.

References

[1] Abubakar S S and Yahaya A 2013 Demonstrating The Invariance Property Of Hotelling T^2 Statistic. *International Journal of Innovative Research in Science, Engineering and Technology*, Vol. 2, Issue 7, July 2013, pp. 2915-2919. http://www.ijirset.com

[2] Aurangzaib and Khan M S 2017 Establishing Relationship of the Variation in Raw Material Quality on Productivity. *IOSR Journal of Business and Management (IOSR-JBM)*, e-ISSN: 2278-487X, p-ISSN: 2319-7668. Volume 19, Issue 8. Ver. V. (August 2017), pp. 12-28. www.iosrjournals.org

[3] Capaci F, Vanhatalo E, Bergquist B, and Kulahci M 2017 Managerial Implications for Improving Continuous Production Processes. *In Proceedings* of The 24th EurOMA conference. pp. 1-11

[4] Dutta S and Genton M G 2014 A non-Gaussian Multivariate Distribution with All Lower-Dimensional Gaussians and Related Families. *Journal of Multivariate Analysis*, 132 (2014), pp. 82–93. http://www.elsevier.com/locate/jmva

[5] Ferrer A 2014 Latent Structures based-Multivariate Statistical Process Control: A Paradigm Shift. *Quality Engineering*. 26(1): pp. 72-91. doi:10.1080/08982112.2013.846093

[6] Haddad F and Alsmadi M K 2018 Improvement of The Hotelling's T^2 Charts Using Robust Location Winsorized One Step M-Estimator (WMOM). *Punjab University Journal of Mathematics,* Vol. 50(1)(2018) pp. 97-112. ISSN 1016-2526

[7] Korkmaz S, Goksuluk D and Zararsiz G 2014 MVN: An R Package for Assessing Multivariate Normality. *The R Journal*, Vol. 6/2, December 2014, pp. 151-162. ISSN 2073-4859

[8] Lee M H 2012 Variable Sampling Rate Hotelling's T^2 Control Chart with Runs Rules. *South African Journal of Industrial Engineering,* May 2012 Vol 23(1): pp 122-129. http://sajie.journal.ac.za

[9] Leoni R C, Costa A F B, Sampaio N A S, Silva J W J and Ribeiro R B 2015 A Geometric Approach to Illustrate the Autocorrelation Effect in T^2 Control Chart of Hotelling. *Applied Mathematics* 2015, 5(2): pp. 39-47. DOI: 10.5923/j.am.20150502.01

[10] Lim S A H, Antony J and Arshed N 2016 A Critical Assessment on SPC Implementation in the UK Food Industry. *Systemics, Cybernetics and Informatics*, Volume 14 - Number 1 - Year 2016, pp. 37-42

[11] Ramzan M B, Kang C W and Sarkar B 2016 A Review Paper On Offline Inspection Of Finished And Semi-Finished Products And Emerging Research Directions. *Yugoslav Journal of Operations Research,* 26 (2016), Number 4, pp. 395-422. DOI: 10.2298/YJOR150815006R

[12] Rogalewicz M 2012 Some Notes on Multivariate Statistical Process Control. *Management and Production Engineering Review,* Volume 3 • Number 4 • December 2012 • pp. 80–86. DOI: 10.2478/v10270-012-0036-7

[13] Yahaya S S S, Ali H and Omar Z 2011 An Alternative Hotelling T^2 Control Chart Based on Minimum Vector Variance (MVV). *Modern Applied Science*, Vol. 5, No. 4; August 2011, pp. 132-151. www.ccsenet.org/mas

[14] Hidayat Y, Purwandari T, Alvionita and Sukono 2016 Analysis Of Accuracy Multivariate Control Chart T^2 Hotelling Free Distribution With Outlier Removal (Case Study: Production Process Ink at PT. EPSON Batam). *Proceedings of the 2016 International Conference on Industrial Engineering and Operations Management,* Detroit, USA, September 23-25, 2016, pp. 547-558. http://ieomsociety.org/ieomdetroit/pdfs/216.pdf

[15] Hidayat Y, Purwandari T, Abdurahman S, Sukono and Bon A T 2018 Product Quality Analysis RS CD 10 SY-C9 Using Multivariate Statistical Process Control. *Proceedings of the International Conference on Industrial Engineering and Operations Management,* Bandung, Indonesia, March 6-8, 2018, pp. 2743-2748. http://ieomsociety.org/ieom2018/papers/649.pdf

[16] Zhan X, Jiang S, Yang Y, Liang J, Shi T and Li X 2015 Inline Measurement of Particle Concentrations in Multicomponent Suspensions using Ultrasonic Sensor and Least Squares Support Vector Machines. *Sensors,* 2015, *15*, pp. 24109-24124; doi:10.3390/s150924109. ISSN 1424-8220. www.mdpi.com/journal/sensors

IOP Conf. Series: Materials Science and Engineering **621** (2019) 012017 doi:10.1088/1757-899X/621/1/012017

Spatiotemporal Dengue Disease Clustering by Means Local Spatiotemporal Moran's Index

I G N M Jaya[1,6], Y Andriyana[2], B Tantular[3], Zulhanif[4], BN Ruchjana[5]

[1,2,3,4)]Statistics Department, Padjadjaran University, Bandung, Indonesia

[5)]Mathematics Department, Padjadjaran University, Bandung, Indonesia

[6)]Faculty of Spatial Science-Groningen University, Netherlands

mindra@unpad.ac.id

Abstract. Spatiotemporal analysis has been used widely to explain some geographic phenomenon, especially in an epidemiology study. Spatial and temporal autocorrelation coefficients are usually used to assess the spatial and temporal dependencies in set geographic events. However, those statistics are usually computed separately and may lead to the misleading conclusion. Analysing spatiotemporal autocorrelation would be useful to understand the geographical evolution simultaneously. Spatiotemporal autocorrelation can be used to identify the spatiotemporal clustering and outlier via local spatiotemporal autocorrelation. This paper develops a method to estimate and test the local spatiotemporal autocorrelation based on the local spatial Moran's Index. Randomization permutation test is used to obtain the p-value which is used to construct the disease clustering. The method was applied to identify the spatiotemporal clustering and outlier detection for dengue disease data in Bandung city. Based on image analysis, this method presents the better result compare than the local spatial Moran's Index which is done for every time separately.

1. Introduction

Dengue disease is an injurious infectious disease transmitted via the female *Aedes-spp.* mosquito from one person to another [1]. Dengue disease is globally endemic in most (sub) tropical countries, and particularly in urban and semi-urban regions. The disease appears to be continuing to spread rapidly. The transmission of pathogens of dengue disease is promoted by multiple risk factors such as environmental and climates factors. Analysing the spatiotemporal pattern of dengue disease may be useful to develop the etiological hypothesis of the disease with regard to environmental or lifestyle-related risk factors [2]. Where the etiology of a disease is not yet well-established, e.g., dengue disease, it may be important to determine whether the cases of the disease are occurring independently or if they seem to be related [3]. In this research, we will be concerned with the condition in which related disease cases occur closely together, both in space and time and probability create the spatiotemporal clusters.

Statistical methods have been widely used for spatial cluster identification in epidemiology study, especially for the processing of geographic data. Those methods are: join-count statistics [4]; Ohno statistics [5]; Poisson statistics [6]; Global Moran's I ([7]; [8]); Global Geary's C ([7]; [8]); General Getis-Ord's G [9]; Local Moran's I ; and Local $Gi(d)$ and $Gi*(d)$ ([9]; [10]; [11]). Statistical approach

for spatial autocorrelation such as the Moran's I and Geary's C methods present a global autocorrelation for overall degree of spatial autocorrelation in a dataset.
However, all of the methods only accommodate the spatial variation which is ignoring the temporal attribute. Ignoring temporal attribute may lead to the misleading conclusion. In this paper, we develop Local spatiotemporal Moran's Index which accommodates the temporal dependencies. The p-value of local spatiotemporal Moran's Index is used to define disease clustering which developed based on randomization permutation test.
The structure of the remainder of this paper is as follows; Section 2 presents the spatiotemporal Moran's Index, Section 3 applies the randomization method to obtain p-value. Section 4 present application for dengue disease data in Bandung City and section 5 presents the discussion.

2. Spatial and spatiotemporal clustering

2.1. Spatial Clustering

Spatial clustering can be seen as a spatial autocorrelation function. The high spatial autocorrelation indicates there is a spatial cluster in the data set. There are two measurement of spatial autocorrelation. Both are global and local spatial autocorrelation. Global and local Moran's Index are commonly used in geographical study to identify the global and local dependencies for a set geographical data. Global autocorrelation measure the spatial dependencies in global way where the autocorrelation coefficient is unique for all regions. The coefficient of global Moran's Index value between -1 and 1. The value closed to positive one indicate there is a spatial cluster. Global Moran's index is based on regression coefficient the observed variables with its lag dependence. Global Moran's I can be defined as [12]; [10]:

$$I = \frac{n\sum_{i=1}^{n}\sum_{j=1}^{n} w_{ij}\,(y_i - \bar{y})(y_j - \bar{y})}{\sum_{i=1}^{n}\sum_{j=1}^{n} w_{ij}\sum_{i}^{n}(y_i - \bar{y})^2} \quad (1)$$

where n denotes the number of spatial units being analysed; y_i is the attribute value of spatial object i; $\bar{y}$ is the mean of all y. Spatial weight matrix w_{ij} denotes the element of spatial weight matrix for observation of spatial unit i and spatial unit j:

$$w_{ij} = \begin{cases} 1\ if\ i\ and\ j\ are\ connected \\ 0\ otherwise \end{cases} \quad (2)$$

The second, is local Moran's Index. This is commonly used to detect the outlier or hotspot in geographical data set. This method is useful for identifying high risk cluster region. The value of local Moran's does not have to be interval -1 to 1. The larger positive value of local Moran's Index indicates there is a strong local spatial cluster. For spatial local Moran's Index can be defined as:

$$I_i = \frac{n(y_i - \bar{y})\sum_{j=1}^{n} w_{ij}\,(y_j - \bar{y})}{\sum_{i}^{n}(y_i - \bar{y})^2} \quad (3)$$

The normal approximation (Z-test) and permutation test are usually used to test the signification of Global and Local Moran's I.

2.2. Spatiotemporal clustering

Spatial Moran's Index can be extended becomes spatiotemporal Moran's Index to accommodate the temporal autocorrelation. Temporal autocorrelation is an important component in evaluating the spatial evolution of the geographical phenomenon. Ignoring temporal attributes may lead to the misleading conclusion because of the number of incidence in location i at time t are connected not only with location j but also time s. The local spatiotemporal autocorrelation can be used to identify clustering of regions and time periods (where and when) simultaneously (spatiotemporal *Hot Spot*).
Here, we develop spatiotemporal Moran's Index based on Autoregressive model order one (AR 1). We assume, the observed variable y_{it} are related to the y_{it-1} and y_{it+1} in another word, $w_{st} = 1\ if\ |s - t| < 1$. Using the Spatiotemporal Moran's Index, it was possible to calculate Global and Localized indices that account for spatial and temporal autocorrelation simultaneously.

The spatiotemporal weight matrix have to be measured in more general way. We only need to re-structured the spatial weight matrix (2) becomes spatiotemporal weight matrix. Spatiotemporal weight have to account for the spatiotemporal autocorrelation between y_{it} and y_{js}, defined as [13]:

$$\widetilde{w}_{(it,js)} = \begin{cases} w_{ij}, & if\ t = s \\ 1, & if\ i = j\ and\ |t-s| = 1 \\ 0, & otherwise \end{cases} \quad (4)$$

where w_{ij} is defined in (2). Hence, the spatiotemporal Global Moran's Index (MoranST) is defined as [13]:

$$MoranST = \frac{nT\sum_{i=1}^{n}\sum_{t=1}^{T}\sum_{j=1}^{n}\sum_{s=1}^{T}\widetilde{w}_{(it,js)}(y_{it}-\bar{y})(y_{js}-\bar{y})}{\sum_{i=1}^{n}\sum_{t=1}^{T}\sum_{j=1}^{n}\sum_{s=1}^{T}\widetilde{w}_{(it,js)}\sum_{i=1}^{n}\sum_{t}^{T}(y_{it}-y)^2} \quad (5)$$

where $\bar{y}$ is the mean of the observed residuals r_{it} over T. MoranST coefficient between -1 to 1. The value closed to 1 corresponds strong positive spatiotemporal autocorrelation, while 0 corresponds to complete spatiotemporal randomness.

For spatiotemporal Local Moran's index is developed based logical thinking of spatial local Moran's Index with substitute the spatial weight matrix by spatiotemporal weight matrix in (4). From (3) the spatial local Moran's index can be written as:

$$MoranST_{it} = \frac{n(y_{it}-\bar{y})\sum_{j=1}^{n}\sum_{s=1}^{T}\widetilde{w}_{(it,js)}(y_{js}-\bar{y})}{\sum_{i=1}^{n}\sum_{t}^{T}(y_{it}-\bar{y})^2} \quad (6)$$

The value of Local spatiotemporal local Moran's I does not have to be interval -1 to 1.

The complexity in developing the spatiotemporal Moran's Index is how to find out the exact statistical testing for it. Spatiotemporal Moran's Index includes spatial and temporal variation which have to accommodate in developing statistical testing. [14] uses simple technique which assumes the statistical test for spatiotemporal Moran's Index is similar with spatial Moran's Index with took the strong assumption that for event i and event j, the spatiotemporal weight is defined as $w_{ij}t_{ij}$. It means, the spatial and temporal component have a similar event.

Spatial weight matrix w_{ij} denotes the element of spatial weight matrix for observation of spatial unit i and spatial unit j;

$$w_{ij} = \begin{cases} 1\ if\ i\ and\ j\ are\ connected \\ 0\ otherwise \end{cases}$$

Temporal weight matrix t_{ij} denote the element of temporal weight matrix for observation of time unit i and time unit j;

$$t_{ij} = \begin{cases} 1\ if\ |t_i - t_j| \le t_\theta \\ 0\ otherwise \end{cases}$$

It means, the temporal index (i,j) have to follow spatial index (i,j) where this assumption is not realistic because we cannot calculate a spatiotemporal weight for $w_{ij}t_{kl}$. Using [14] assumptions, we cannot take into account the pair of observations y_{it} and y_{js}, we only use observations y_{ii} and y_{jj}. It means we only take into account observation at location i and time i and observation at location j time j. Here we developed a permutation test of spatiotemporal autocorrelation.

2.3. Random Permutation Test of Spatiotemporal Autocorrelation

Random permutation test becomes a popular testing procedure in hypothesis testing because it does not rely need assumption about the distribution of the data. In geographical analysis, the normality assumption is hard to be satisfied. The basic idea of the random permutation test is a central limit theorem. The p-value is used in testing obtained by resampling on the observed data form many times, where p-value presents the probability of obtaining data as extreme as the observed data when the null hypothesis is true [15]. If the data are drawn many times with replacement in case the null hypothesis being true, the number of cases with data as extreme as the observed data could be counted, and a p-value calculated. The steps of the random permutation test for spatiotemporal autocorrelation are as follow [15] :

1. Define spatiotemporal weight matrix as in (4)
2. Let $y^0 = (y_{11}, y_{12}, \dots, y_{nT})$ denotes the vector observed spatiotemporal data and construct the corresponding value of spatiotemporal Global or Local Moran's I, $f^0 = f(y^0)$ where f can be a function of Global and Local spatiotemporal Moran's I
3. Simulate M random permutations, $p^j = (p_1^j, \dots, p_{nT}^j)$ of integer $(1,2,..,nT)$
4. For each permutation, p^j, construct the corresponding permuted data vector, $y(p^j) = \left(y_{p_1^j}, \dots, y_{p_{nT}^j}\right)$ and resulting f, denote by $f^j = f[y(p^j)], j = 1,2, \dots, M$
5. Rank the values $(f^0, f^1 \dots, f^M)$ from low to high, so that if f^j is the kth highest value then $rank(f^j) = k$
6. If $rank(R^0) = k$ then construct p-value for this test to be
$$\alpha = \frac{k}{M+1} \tag{7}$$
7. Decision for testing
   (i) If α is low (e.g., $\alpha \leq 0.05$) indicates there is significant positive spatiotemporal autocorrelation.
   (ii) Inversely, if α is high (e.g., $\alpha \geq 0.95$) indicates there is significantly negative spatiotemporal autocorrelation
   (iii) If neither (i) or (ii) satisfied, indicate the spatiotemporal independence hypothesis, H_0, cannot be rejected.

3. Application: Dengue Disease in Bandung City

Bandung is the most economically advanced city in West Java. The number of dengue incidence usually high for every year. In this section, to illustrate the Moran's Index for spatiotemporal data, dengue data in 30 Bandung regions over the period January-December, 2016 have been used. In 2016 there is 3880 number of dengue cases in Bandung. Coblong, Buah Batu, and Rancasari are the three sub-districts with the highest number of cases. The number of incidences is increasing from January to May, and the highest dengue incidence in May. Boxplot present there are some spatial and time have the highest incidence which can be diagnostic as the outlier.

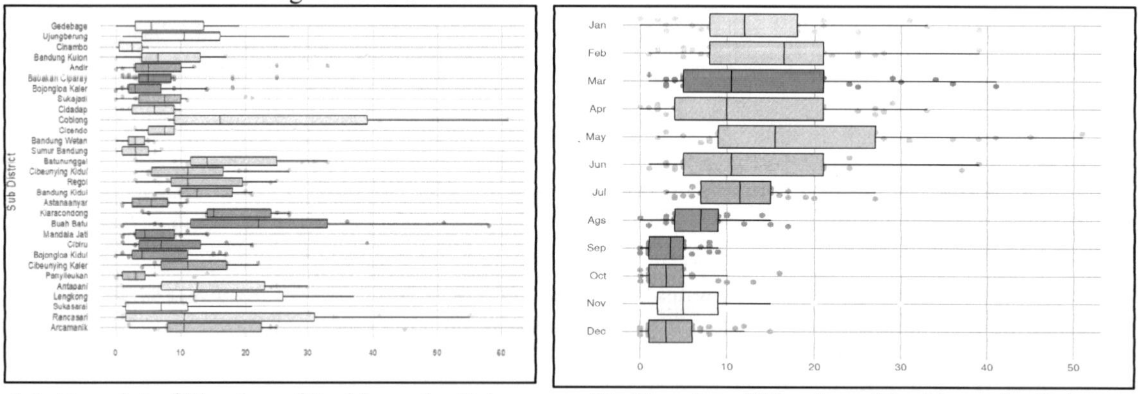

(a) Boxplot of Number of Incidence by Sub District

(b) Boxplot of Number of Incidence by Sub District

Figure 1. Distribution of Dengue disease in Bandung, 2016

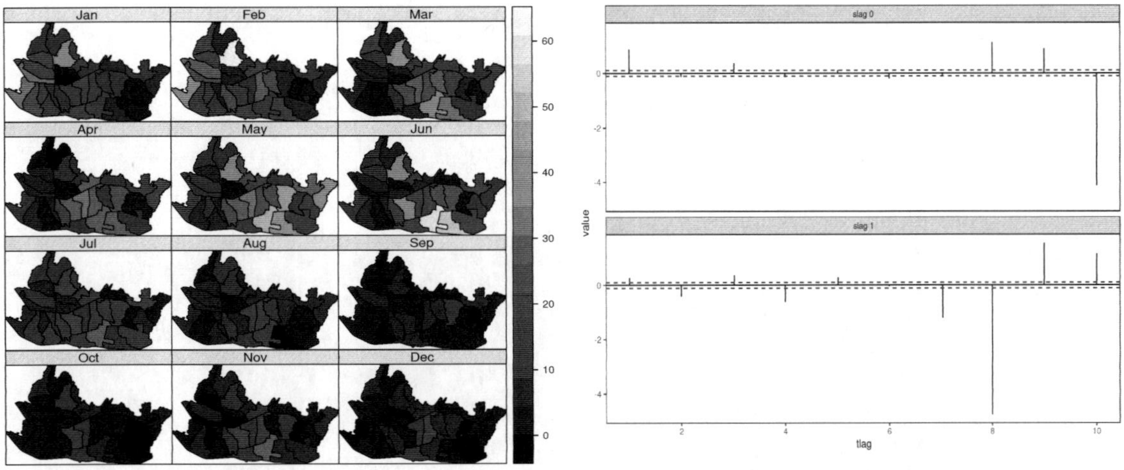

(a) Spatiotemporal Distributio of Dengue Incidence in Bandung

(b) Partial Spatiotemporal Autocorrelation Function

Figure 2. (a) Spatiotemporal Distirbution of Dengue diseas in Bandung, 2016 and (b) Partial Spatitomporal Autocorrelation Function of dengue indicidence

Spatiotemporal distribution of dengue disease shows that the number incidence of dengue disease is different for every location and time. Figure 2(b) present there is a temporal autocorrelation in the data which are several time lags are significant.

Table 1. Global Spatial Autocorrelation for eachtime unit

Month	Moran's I	p-value
Jan	0.303	0.005**
Feb	0.095	0.115
Mar	0.301	0.005**
Apr	0.385	0.002**
May	0.295	0.003**
Jun	0.319	0.005**
Jul	0.072	0.175
Aug	0.219	0.018*
Sep	0.001	0.358
Oct	0.300	0.005**
Nov	0.183	0.037*
Dec	0.236	0.018**

Note:
** significant at level $\alpha < 0.01$
* significant at level $\mathbf{0.01 < \alpha < 0.05}$

IOP Conf. Series: Materials Science and Engineering **621** (2019) 012017 doi:10.1088/1757-899X/621/1/012017

Table 2. Global Temporal Autocorrelation for each spatial unit

District	Moran's I	p-value
Gedebage	0.659	0.005**
Ujungberung	0.737	0.001**
Cinambo	0.441	0.006**
Bandung Kulon	0.451	0.039*
Andir	0.453	0.034*
Babakan Ciparay	0.406	0.044*
Bojongloa Kaler	0.385	0.046*
Sukajadi	0.622	0.004**
Cidadap	0.079	0.291
Coblong	0.604	0.010*
Cicendo	0.298	0.084*
Bandung Wetan	0.205	0.145
Sumur Bandung	0.307	0.104
Batununggal	0.628	0.004**
Cibeunying Kidul	0.383	0.038*
Regol	0.443	0.022*
Bandung Kidul	0.404	0.027*
Astanaanyar	0.292	0.102
Kiaracondong	0.719	0.001**
Buah Batu	0.472	0.020*
Mandala Jati	0.591	0.008**
Cibiru	0.399	0.024*
Bojongloa Kidul	0.268	0.104
Cibeunying Kaler	0.484	0.032*
Panyileukan	0.526	0.018*
Antapani	0.580	0.008**
Lengkong	0.459	0.027*
Sukasari	0.480	0.026*
Rancasari	0.639	0.005**
Arcamanik	0.604	0.002**

Table 1 present for several months the spatial Moran's Index are significant at level 0.05. Table 2 present the temporal Moran's Index for 30 sub-districts in Bandung. Several sub-districts have high values of Moran's Index. We also compute the Local spatial Moran's Index by month as presented in Figure 3.

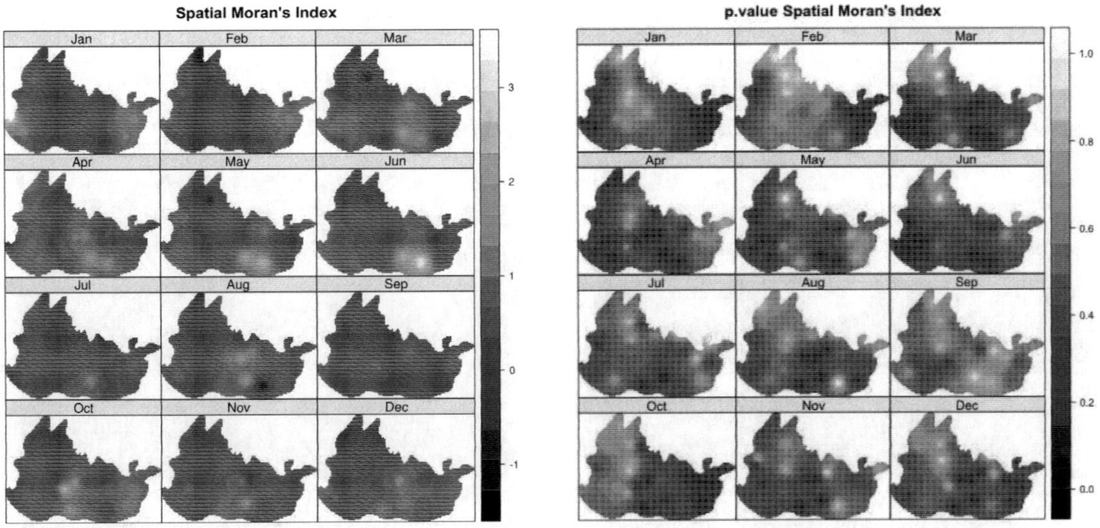

(a) Spatial Moran's Index (b) p-value Spatial Moran's Index

Figure 3. Local Spatial Moran's Index for Dengue Disease in Bandung

Figure 3 shows the local spatial Moran's index which is computing separately by month. However, the Moran's statistics fail to identify temporal cluster and the outlier in Coblong and Buah Batu sub-districts. The figure could be present the clear spatiotemporal clustering and outlier. This is because of ignoring the temporal dependencies. From that information it is important to estimate the spatiotemporal Moran's Index simultaneously.

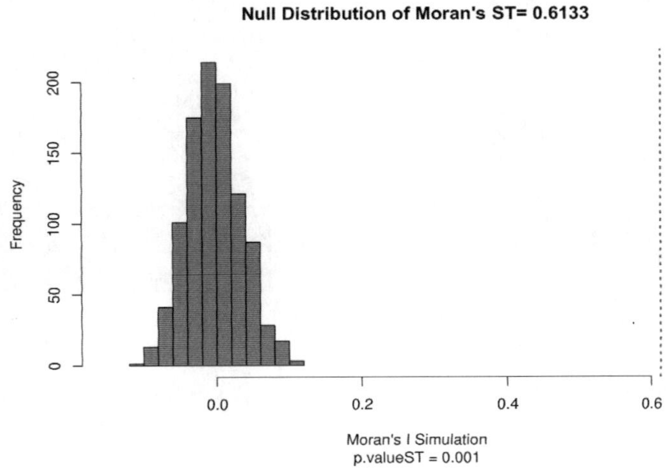

Figure 4. Spatiotemporal Moran's Index with Permutation Test

The dengue incidence data in Bandung city has high spatiotemporal Moran's Index coefficient, 0.6113 with small p-value, 0.001 (see Figure 4). It means there is a spatiotemporal dependency in the data. The observation in location i and time t is related to the neighbouring locatin at time $t-1$. We also presented the Spatiotmeporal Local Moran's Index in isopleth maps. The map present the spatiotmeporal cluste with high incidence and the outlier detection.

IOP Conf. Series: Materials Science and Engineering **621** (2019) 012017 doi:10.1088/1757-899X/621/1/012017

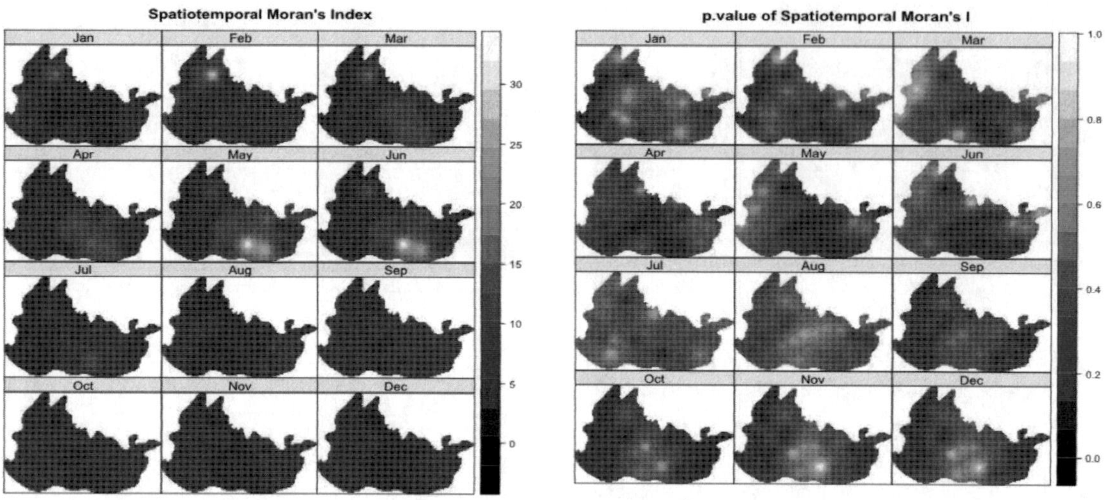

(a) Spatiotemporal Moran's Index (b) p-value of Spatiotemporal Moran's Index

Figure 5. Spatiotemporal Moran's Index for dengue disease in Bandung city

The districts with high incidence are Coblong, Buah Batu and Rancasari. Figure 5(a), Coblong is an outlier because there only one district in north of Bandung has high incidence. However, Buah batu and Rancasari create a cluster for high incidence district at May and June. This result present that spatiotemporal Moran's Index is a valuable tool to identify the spatiotemporal cluster and outlier.

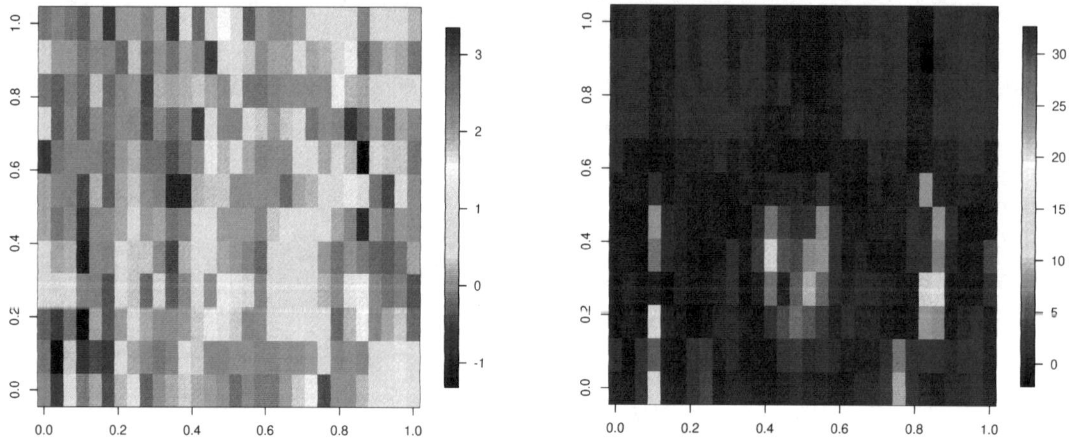

(a) Image Analysis of Matrix Spatial Moran's Index (b) Image Analysis of Matrix p-value Spatial Moran's Index

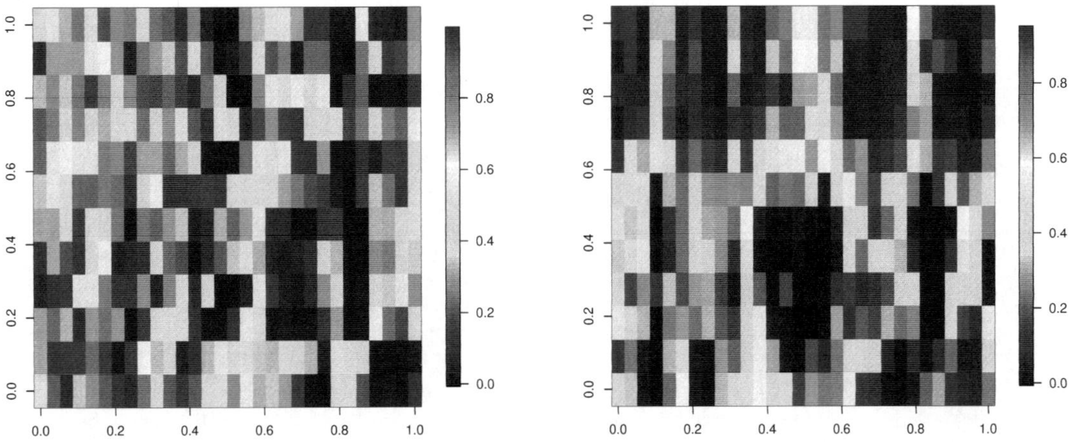

(c) Image Analysis of Matrix Spatial Moran's Index

(d) Image Analysis of Matrix p-value Spatial Moran's Index

Figure 6. Comparison of Spatial Moran's Index and Spatiotemporal Moran's Index

We compare the spatial Moran's index which is computed by months separately and spatiotemporal Moran's Index. We can see there is a significant difference between both statistics. Spatiotemporal Moran's Index present clearly spatiotemporal cluster and outlier. Here we present in detail the spatiotemporal clustering based on number of dengue cases in Bandung city.

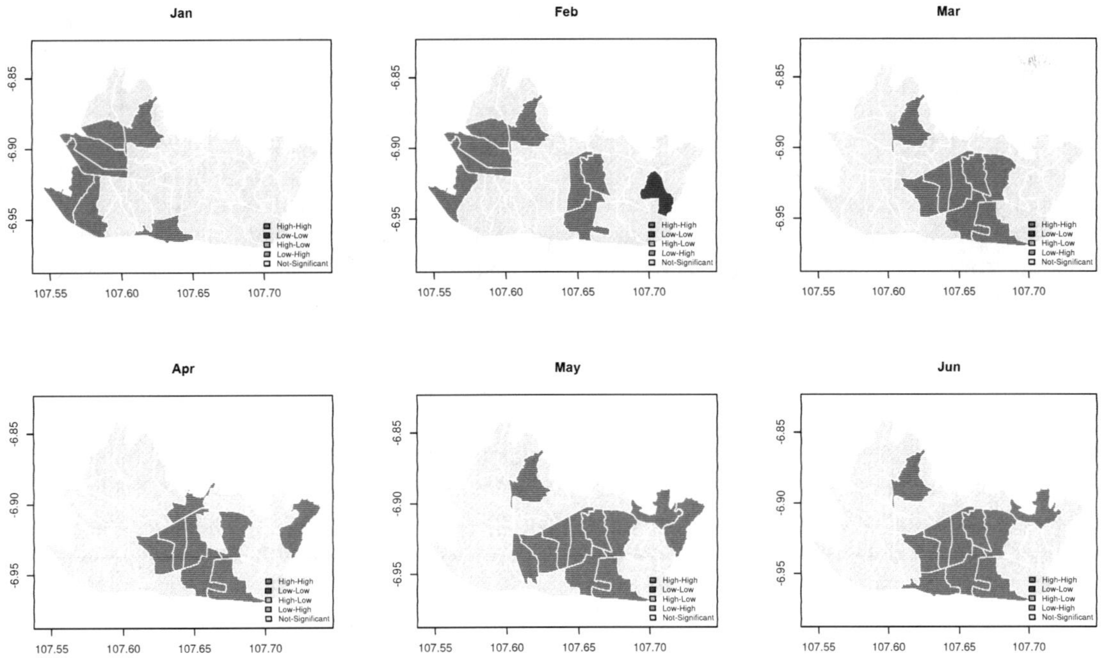

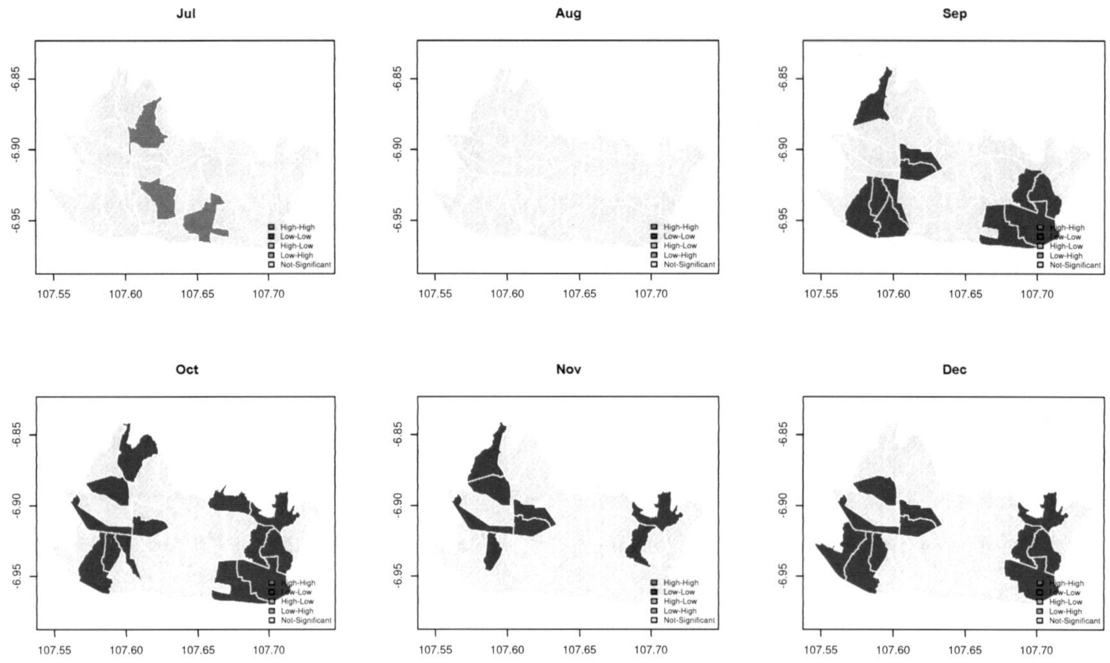

Figure 7. Spatiotemporal clustering of dengue disease in Bandung city, 2016 by means local spatiotemporal Moran's I

Figure 7 displays Spatiotemporal clustering of dengue disease in Bandung city, 2016 by means local spatiotemporal Moran's I. According to the distribution of the number of cases pattern in Figure 2 and spatiotemporal autocorrelation in Figure 5, using permutation test we obtain the spatiotemporal clustering as a group of significant spatiotemporal autocorrelation. There are five different labels (1) high-high (2) low-low (3) high-low (4) low-high and (5) not significant. Our method presents clearly that the high-risk region with large number incidence found in January up to July. Those months have a number of incidences larger than the other month. The cluster of the high-risk region change over time. This result indicates that there is temporal characteristics that change the clustering structure by time for example there a change in environmental factors and weather. This information may be very useful information for the health department in an effort to overcome dengue disease problem in Bandung city through understanding the distribution patterns of disease diseases - where and when high case rates will be found.

4. Discussion

Temporal autocorrelation is important component in spatiotemporal analysis of geographical data set. Ignoring temporal component might be lead to the misleading result. In order to identify the spatiotemporal cluster of dengue disease data, we develop a local spatiotemporal autocorrelation where the cluster can be defined as grouping regions with significant local spatiotemporal autocorrelation. Here we used permutation test to test the significant of global and local spatiotemporal Moran's I. We developed R package which can be used to compute Global and Local spatial, temporal and spatiotemporal Moran's I for Autoregressive Oder 1 model and can be developed for high order of AR. We compare the spatial and Spatiotemporal Moran's Index which produce very different results. Image analysis shows clearly that the spatial Moran's Index for each time, and spatiotemporal Moran's Index present different cluster and outlier. The spatial Moran's Index for each time periods produces under estimate spatiotemporal autocorrelation compare than spatiotemporal Moran's Index which computed simultaneously. The spatial Moran's Index also fails to detect the outlier. As we see from the original data, sub-district Coblong has highest number incidence of dengue disease which is different from the others sub-districts. For further research, the relative risk estimate could be combined with local spatiotemporal Moran's Index to construct map with clear pattern of geographical disease evolution and disease clustering. The complete approach of disease mapping can be seen in [16].

Acknowledgement
This paper is funded by RFU Unpad contract: 1732 d/UN6.RKT/LT/2018. The authors thank Rector Universitas Padjadjaran and to the anonymous referee whose valuable checking has improved this paper.

References

[1] Acharya K B, Cao C, Lakes T, Chen W and Naeem S 2016 *BMC Public Health* **16** 1
[2] Lione G and Gonthier P 2016 *Analytical and Theoretical Plant Pathology* **106** 19
[3] Mantel N 1967 *Cancer Res* **27** 209
[4] Upton G J and Fingleton B 1985 *Spatial data analysis by example* vol 1 (United Kingdom: John Wiley & Sons Ltd)
[5] Ohno Y, Aoki K and Aoki N 1979 *Int J Epidemiol* **8** 273
[6] Miyawaki N and Chen S C 1981 *Social Science & Medicine Part D: Medical Geography* **15** 93
[7] Cressie N 1993 *Statistics for Spatial Data* (New York: Wiley-Interscience)
[8] Legendre P and Legendre L 1998 *Numerical Ecology* (Amsterdam: Elsevier)
[9] Getis A and Ord J 1992 *Geographical Analysis* **24** 189
[10] Anselin L 1995 *Geographical Analysis* **27** 93
[11] Ord J and Getis A 1995 *Geographical Analysis* **27** 286
[12] Chen Y 2013 *PLOS ONE* **8** 1
[13] Anderson C and Ryan L M 2017 *Int J Environ Res Public Health.* **14** 1
[14] Lee J and Li S 2016 *Geographical Analysis* **1** 36
[15] Good P 2005 *Permutation, Parametric and Bootstrap Tests of Hypotheses* (USA: Springer)
[16] Jaya I G N M, Folmer H, Ruchjana B N, Andriaya Y and Kristianti F 2017 *Regional research frontiers* vol 2 (USA: Springer)

IOP Conf. Series: Materials Science and Engineering **621** (2019) 012018 doi:10.1088/1757-899X/621/1/012018

Numerical Comparison of some Crossover in Genetic Algorithm for Two Dimensional Global Minimization Problem using C++

Muhamad Deni Johansyah[1], Herlina Napitupulu[2*], Eddy Djauhari[3], Sukono[4], and Julita Nahar[5]

[1,2,3,4,5]Department of Mathematics, FMIPA, Universitas Padjadjaran, Jatinangor 45363, Sumedang, Jawa-Barat, Indonesia

Corresponding author's email: *napitupuluherlina@gmail.com

Abstract. In this paper we simulate the comparison of numerical results of the different crossover method in genetics algorithm for solving global minimization problem. In this paper, the objective function is for 2 two variables function. The genetics algorithm is run using C++ program and implemented to several benchmark test functions of global optimization. The performance of each crossover method is analyse according to the numerical results, which show that certain crossover method is better for certain benchmark test function.

1. Introduction

Genetic algorithms (GA) was found by John Holland and developed by David Golberg. GA is numerical optimization algorithms which its idea is from natural selection and natural genetics [3] which is regarded as a heuristic approach. This method is known as direct, parallel, as well as powerful technique in optimization methods. This method is able to optimize continuous or discrete variables, with non-derivative needed, simultaneously searches from a large domain of objective function, deals with a large number of variables, optimizes variables with extremely complex objective function, provides several optimum solutions instead of one and so on [1,4,5]. This method is applicable to various fields of optimization case. This makes GA being an attractive method to be studied [6]. Some literature are discussing the genetics algorithm for global optimization purpose such as: in [7] to achieve the global optimization the authors use rastrigin function as illustration, in [8] shows the implementation of genetics algorithm optimization to control non-linear direct torque control of induction motor drive, and in [9] the research is made to optimize the surface roughness prediction model using Genetic Algorithms (GA) to optimize the objective function.

2. Methods

One of the stochastic methods to solve the minimization problems is genetics algorithm method. Our goal is to solve the minimization problem using Genetic Algorithm method. The general genetics algorithm pseudo code, suggested by Golberg, is given below and its implementations are given in Section 3 for solving minimization problem.

IOP Conf. Series: Materials Science and Engineering **621** (2019) 012018 doi:10.1088/1757-899X/621/1/012018

Pseudo code of Genetics Algorithm
Step 1. Initialization
Step 2. Encoding
Step 3. Generate the Initial population
Step 4. Chromosome evaluation (Calculate and Evaluate of fitness value)
Step 5. Calculation of pk; $cg = cg + 1$,
    Step 5.1. if $pk \geq \theta$ or $cg \geq jg \geq \theta$ then decoding, go to Step 6
    Step 5.2. else
        Step 5.2.1. Selection (Individual Selection)
        Step 5.2.2. Crossover
        Step 5.2.3. Mutation
        Step 5.2.4. Create a New Population and go to Step 4
Step 6. Termination

3. Genetic Algorithm in Global Optimization

Let $\mathbb{R}^n$ the n −dimensional real Euclidean space, $x = (x_1, x_2, .., x_n) \in \mathbb{R}^n$, and let $f: \mathbb{R} \to \mathbb{R}^n$ be a real-valued function. In this paper the objective is to find the global minimizers of f, that is, the points $x^* \in \mathbb{R}^n$ such that $f(x^*) \leq f(x)$ for all $x \in \mathbb{R}^n$. We restrict the scope only for $n = 2$.

To construct the GA program to solve the minimization problem using, we use the following criteria.

- The objective function is minimizing the function f with
- Fitness that we calculate is for minimization;
  N = population size. To compare the performance we take several values of N in this paper
- The total fitness calculated by: $F = \sum_{k=1}^{N} eval(v_k), k = 1,2, .., N$
- The formula to calculate the cumulative probability q_k for each chromosome v_k is given by $q_k = \sum_{k=1}^{N} p_k, k = 1,2, .., N$.

The genetics algorithm processes generally separated into following important steps: encoding, selection, crossover, and mutation ([6]). The details explanation of each steps is given in the following..

Pre-step 1: Initialization

The initialization is done as follows, we initialized the number of generation jg, the counter generation cg, the number of population pop, the crossover probability pc, and number of mutation per generation mut, and the threshold θ.

Pre-step 2: Encoding

The objective is to find $x = (x_1, x_2)$ which minimize of $f(x)$ in the interval solution

$$x_1^L \leq x \leq x_1^U, x_2^L \leq x \leq x_2^U.$$

In this research we use binary encoding method to represent chromosome and the gen length (denoted by m_j). We calculate m_j ($j = 1,2$) as

$$2^{m_j - 1} < (x_j^U - x_j^L) * 10^k < 2^{m_j} - 1.$$

To convert the binary to decimal for variable x, we use formula

$$x_j = x_j^L + decimal(substring_j) * \frac{(x_j^U - x_j^L)}{2^{m_j - 1}}. \quad (1)$$

For each case problem we have $m_1 = m_2$, so that we will have length of chromosome or number of gen $m = m_1 + m_2$.

Step 1: Initial Population

A set of N initial population is obtained randomly. We use the $N = 300, 600,$ and 800 random initial population, which are in the binary form of length bit. The decimal value of each initial population is calculated by substituting the decimal of each substring to formula (1).

Step 2: Chromosome Evaluation
In our case, the fitness value of chromosome is minimization problem, is the value of objective function f. The fitness value of each chromosome is evaluated in this step, in order to see the best value of f (minimum value).

Step 3: Calculation of Population Convergence Percentage
The Population Convergence Percentage (pk) defined as percentage of total number of individual with most total number of the same fitness value. It use the formula

$$pk = \frac{n}{pop} * 100\%$$

where n is the most total number with similar fitness value with pop is the number of individual. The calculation of counter generation is given by formula,

$$cg = cg + 1.$$

Step 4: Termination Criteria
The termination holds when the counter generation cg is reaching the number of generation ($jg = 1000$) or the convergence of population pk reaching the threshold level ($\theta = 90\%$). After population is initialized, the value of $cg = 0$. If there is no such individual with similar fitness number, then $=$ 10% , means that that the genetics algorithm process will continue.

Step 5: Chromosome Selection
To calculate the fitness value of minimization problem, the following formula is used:

$$eval(v_k) = \frac{1}{f(x^G) - f(x)},$$

with x^G as the global minimizer of f. The total fitness is obtained by the formula

$$F = \sum_{k=1}^{N} eval(v_k)$$

thus the probability selection p_i for each chromosome v_k is calculated using

$$p_k = \frac{eval(v_k)}{F}.$$

The cumulative probability is calculated by

$$q_k = \sum_{i=1}^{k} p_i.$$

By selecting 10 times the set of 10 chromosome using roulette wheel, the random numbers $r \in [0,1]$ is obtained. We apply the following criteria to addresses the position to replace the selected $k - th$ current individual,

$$if \begin{cases} q_{k-1} < r \leq q_k, & select \quad v_k \\ r < q_1, & select \quad v_1 \end{cases}.$$

Step 6: Crossover
The crossover will be done to the new population; in this paper the research is to compare the performance of the Crossover method as follows ([6]):

- Single Point Crossover (SPX). SPX is the very simple crossover by generate one random number (the value is the number from 1 to gen length of one chromosome) as the cutting point. The algorithm of SPX is as follows [5]:
  a) Generate one random number (in the interval $[1, gen\ length]$), this number will be the initial position of gen of the chromosome to be crossover
  b) Switch the allele which is noted by parent 1 and parent 2.

- Two Point Crossover (TPX). TPX is the crossover with two cutting point, where two random number are generated (the value is the number from 1 to gen length of one chromosome) as cutting points. The algorithm of TPX is as follows [5]:
  a) Generate two random number (in the interval $[1, gen\ length]$), in which this numbers will be used for as the beginning and ending position respectively of the chromosome to be crossover.
  b) Switch the allele which is noted by parent 1 and 2.

Step 7: Mutation
Number of gen in one generation is $m * N$. If we want to arrange the mutation probability value equal to pm, then we should have calculate number of mutation (nm) as

$$nm = m * N * pm$$

In this simulation, we use $pm = 0.001$, then, in the mutation process we generate randomly $nm = 4$ number of integer in interval $[1, pop * m]$. This 4 number is used as location in which gen will be mutated. The replacement process of allele is 1 and 0, as follows

$$allele \begin{cases} 0 \to 1 \\ 1 \to 0 \end{cases}.$$

After the mutation process then the calculation in one generation is finished. The next process is to repeat the process in step 2 of the algorithm until the termination criteria is satisfied.

Step 8: Decoding
Decoding is the process to recode the gens in a chromosome so that the value is return into the value (before encoded).

4. Numerical Results and Discussion
SPX and TPX crossover methods in genetics algorithm are compared computationally. The programming is using C++ Microsoft Visual Studio 2010, which is applied to following 10 benchmark testing function of global optimization.

We initialized the number of generation $jg = 1000$, the counter generation $cg = -1$, the number of population $pop = 300, 600,$ and 800. The crossover probability $pc = 0.25$, number of mutation per generation $mut = 4$, and the threshold $\theta = 90\%$.

The implementation of crossover method on global optimization in this research is as follows. For SPX method, we generate one random number r_c which its value is from 1 to $gen\ length$ that will be pointing the initial gen position to be crossover in both parents, thus the allele between parents are switched. The $gen\ length = m$ of each problem is based on the calculation of $m_1 + m_2 = m$ as given in step 2. For the TPX method the number of generated random number are two, with value in between 1 to $gen\ length$. Thus this numbers are used as the starting and ending position of chromosome to be crossover, thus the pointed allele are switched between parent 1 and 2.

Problem 1.
Six Hump Back Camel Function has two global minimum solutions, i.e. $x^* = (0.08984, 0.71265)$ and $x^* = (-0.08984, -0.71265)$ where $f(x^*) = -1.03163$.

$$f(x_1, x_2) = 4x_1^2 - 2.1x_1^4 + \frac{1}{3}x_1^6 - x_1x_2 - 4x_2^2 + 4x_2^4$$
$$s.t. -3 \le x_1, x_2 \le 3$$

Problem 2.
Three Hump Back Camel Function has one global minimum solution, i.e. $x^* = (0,0)$ with $f(x^*) = 0$

$$f(x_1, x_2) = 2x_1^2 - 1.05x_1^4 + \frac{1}{6}x_1^6 - x_1x_2 + x_2^2$$

$$s.t. -3 \leq x_1, x_2 \leq 3$$

Problem 3.
Shubert Function I has 760 minima with 18 global minimizers where $f(x^*) = -186.731$

$$f(x_1, x_2) = \{\sum_{j=1}^{5} i\cos[(i+1)x_1 + i]\}\{\sum_{j=1}^{5} i\cos[(i+1)x_2 + i]\}$$
$$s.t. -10 \leq x_1, x_2 \leq 10$$

Problem 4.
Shubert Function II has 760 minima with global minimizers $x^* = (-1.42513, -0.800321)$ where $f(x^*) = -186.731$

$$f(x_1, x_2) = \{\sum_{j=1}^{5} i\cos[(i+1)x_1 + i]\}\{\sum_{j=1}^{5} i\cos[(i+1)x_2 + i]\} + \frac{1}{2}((x_1 + 1.42513)^2 + (x_2 + 0.80032)^2)$$
$$s.t. -10 \leq x_1, x_2 \leq 10$$

Problem 5.
Shubert Function III has 760 minima with global minimizers $x^* = (-1.42513, -0.800321)$ where $f(x^*) = -186.731$

$$f(x_1, x_2) = \{\sum_{j=1}^{5} i\cos[(i+1)x_1 + i]\}\{\sum_{j=1}^{5} i\cos[(i+1)x_2 + i]\} + (x_1 + 1.42513)^2 + (x_2 + 0.80032)^2$$
$$s.t. -10 \leq x_1, x_2 \leq 10$$

Problem 6.
Goldstein-Price Function has 4 minimizers, and 1 among them is the global one with the value $f(0, -1) = 3$. Formula of Goldstein-Price function and its domain where minimizers exist are as follows

$$f(x_1, x_2) = \left(1 + (x_1 + x_2 + 1)^2 \times (19 - 14x_1 + 3x_1^2 - 14x_2 + 6x_1x_2 + 3x_2^2)\right) \times \left(30 + (2x_1 - 3x_2)^2 \times (18 - 32x_1 + 12x_1^2 + 48x_2 - 36x_1x_2 + 27x_2^2)\right)$$
$$s.t. -3 \leq x_1, x_2 \leq 3$$

Problem 7.
Rastrigin function has one global minimizer. It has about 50 minimizers. This typical test of non-linear multimodal function is rather difficult due to its large search space and large number of local minimizers. Its global minimum is $f(0,0) = -2$.

$$f(x_1, x_2) = x_1^2 + x_2^2 - \cos(18x_1) - \cos(18x_1)$$
$$s.t. -2 \leq x_1, x_2 \leq 2$$

Problem 8.
Treccani function has two symmetric global minimizer with the value $f(-2,0) = f(0,0) = 0$. The formula and its domain are as follows

$$f(x_1, x_2) = x_1^4 + 4x_1^2 + 4x_1^2 + x_2^2$$
$$s.t. -2 \leq x_1, x_2 \leq 2$$

Problem 9.
Two Dimensional Sin Square function is a function which exhibits many local minimizers and has one global minimizer with minimum value $f(1,1) = 0$. The formula and domain of this function is as follows,

$$f(x_1, x_2) = \frac{\pi}{2}(10\sin^2(\pi x_1) + [(x_1 - 1)^2(1 + 10\sin^2(\pi x_2))] + (x_2 - 1)^2)$$
$$s.t. -10 \leq x_1, x_2 \leq 10$$

Problem 10.
Himmelblause function has four global minimizer $(3,2), (3.58443, -1.84812), (-2.80512, 3.13131)$, and $(-3.77931, -3.28319)$ where $f(x^*) = 0$. The formula and its domain is given as follows

$$f(x_1, x_2) = (x_1^2 + x_2 - 11)^2 + (x_1 + x_2^2 - 7)^2$$
$$s.t. -6 \leq x_1, x_2 \leq 6$$

We assume that the algorithm is a success in obtained global minimizer if the termination point which converge to x^k satisfying $|f(x^k) - f(x^*)| \leq 10^{-2}$. Each test problem are done about 20 times of trials. The numerical results are given in the Table 1 with the following symbols explanation.

Npop = number of initial points/ initial populations
Time = average elapsed time until termination of program
Iter = average number of iteration until program terminated
ns = number of success in reaching the global minimizer
nf = number of failed in reaching the global minimizer

Table 1. Numerical results of Problem 1-10 for 20 time trials

		SPX				*TPX*			
No.	***Npop***	***Time*** (second)	***Iter***	***ns***	***nf***	***Time*** (second)	***Iter***	***ns***	***nf***
1	300	4.04495	270.4	13	7	2.8349	298.65	14	6
	600	6.66255	344.85	18	2	4.2836	164.5	17	3
	800	9.22085	310.35	15	5	5.8303	188.8	18	2
2	300	2.99485	328.85	11	9	2.64285	282.4	14	6
	600	4.4118	197.25	19	1	6.5925	338.25	14	6
	800	6.2863	201.8	16	4	5.087	156.8	18	2
3	300	3.353	375.05	0	20	3.2943	374.7	0	20
	600	11.7184	584.4	1	19	8.4649	341.85	3	17
	800	10.7398	345.2	1	19	15.39225	591.35	0	20
4	300	6.4954	833	0	20	6.37355	787.75	0	20
	600	11.9507	664.1	1	19	10.9877	618.7	0	20
	800	21.03695	845.6	0	20	19.01645	698.6	1	19
5	300	5.75945	694.35	0	20	6.2136	745.55	0	20
	600	19.0516	882.25	0	20	12.6888	701.2	1	19
	800	21.1309	830.15	1	19	15.94935	600.7	3	17
6	300	4.4139	540.1	0	20	3.7866	442.8	3	17
	600	11.58655	704.95	1	19	7.08345	365.6	6	14
	800	9.72575	390.65	5	15	11.70215	439.1	6	14
7	300	3.0973	349.95	7	13	2.87425	300.25	9	11
	600	6.5634	291.45	11	9	5.8313	256.35	13	7
	800	7.6415	209.85	12	8	13.2184	448.8	11	9
8	300	2.7348	284.7	16	4	2.65685	267.95	16	4
	600	6.27285	281.4	19	1	6.7896	313.2	19	1
	800	7.39475	194.05	20	0	9.5255	340.9	20	0
9	300	5.5345	661.25	1	19	4.40175	505.3	3	17
	600	13.1879	724.15	4	16	9.93855	573.25	7	13
	800	18.31975	708.5	6	14	19.7462	764.4	7	13
10	300	4.12935	484.3	2	18	4.0643	462.5	3	17
	600	9.3262	444.4	6	14	9.1094	418.95	7	13
	800	11.8446	405.8	8	12	12.44535	405.15	7	13

According to the numerical results in Table 1, by comparing between SPX and TPX obtained the following summary. Based on the average time elapsed, Problem 1, 2, 4, 6, 7, 8, 9 and 10 for the 20 times trials the average of the least elapsed time is the procedure using TPX with *Npop*=300. But for Problem 3 and 5, the average of the least elapsed time is the procedure using TPX with *Npop*=600, since for problem 3 and 5, weather the SPX or TPX with *Npop*=300 failed in reaching any global minimum.

Based on the average number of iteration, Problem 1, 3, 4, 6 have the least number of iteration on TPX with *Npop*=600. Thus, for Problem 2, 5, 10 on TPX with *Npop*=800, Problem 7 and 8 on SPX with *Npop*=800, and Problem 9 on TPX with *Npop*=300.

Based on the Average Number of success/failed in obtained global minimum, Problem 1 on SPX with *Npop*=600 and TPX with *Npop*=800, Problem 2 on SPX with *Npop*=600. Problem 3 and 7 on TPX with *Npop*=600. Problem 4 on SPX with *Npop*=600 and TPX with *Npop*=800. Problem 5 on TPX with

Npop=800. Problem 6 on TPX with *Npop*=600 and 800. Problem 8 on SPX and TPX respectively with *Npop*=800. Problem 9 on TPX with *Npop*=600 and 800. Problem 10 on SPX with *Npop*=800.

Generally, by comparing between SPX and TPX based on: the least average elapsed time until termination, the number of success, the least number of iteration for the given benchmark test function with given condition then the TPX method is better than SPX method.

5. Conclusion

We can conclude that for most of the benchmark test functions genetics algorithm with SPX and TPX crossover method is success in reaching the global minimum solution. Generally, by comparing the average elapsed time, the least for all problems is reached by TPX crossover method. According to the number of success, TPX is more than SPX method. For the whole benchmark test function, the least iteration is achieved in by TPX crossover method. It is clear that generally for the given conditions and the given benchmark test function, the TPX method is better than the SPX method.

References

[1] Haupt R L and Haupt S E 2004 *Practical Genetic Algorithms (2nd ed.)* (New Jersey: A John Wiley & Sons, Inc., Publication)

[2] Bimlendu S, Sujata D, Abhinav M, Vinay K, Prasanna K C 2016 A Review over Genetic Algorithm and Application of Wireless Network System *Procedia Computer Science* **78** pp 431-438

[3] Wagner F S, Claudio M N A, Pereira, Pius P M S, Roberto S 2002 *Genetic algorithms applied to turbine extraction optimization of a pressurized-water reactor.* Applied Energy **773** pp 217-222

[4] Kramer O 2017 *Studies in Computational Intelligence Volume 679: Genetic Algorithm Essensials* (Oldenburg: Springer)

[5] Weise T 2009 *Global Optimization Algorithms Theory and Application* (http://www.it-weise.de/, downloaded 10/10/2017)

[6] Adriana F and Fadly R L 2015 *Algoritma Genetika dan Penerapannya* (Yogyakarta: Teknosain)

[7] Bajpai P and Kumar M 2010 *Genetic Algorithm- An Approach to Solve Global Optimization Problem* Indian Journal of Computer Science and Engineering **1** (3) pp199-206

[8] Malhotra R, Singh N, and Singh Y 2011 *Genetic Algorithms: Concepts, Design for Optimization of Process Controllers* **4** (2) Computer and Information Science

[9] Suresh P V S, Rao P V, Deshmukh SG 2002 *A genetic algorithm approach for optimization of surface roughness prediction model.* International Journal of Machine Tools and Manufacture **42** 6 pp 675-680

Acknowledgment

The authors thank to the Universitas Padjadjaran for the grant research funding through the scheme of Research Fundamental UNPAD (RFU) with contract number 2335/UN6.D/KS/2018

IOP Conf. Series: Materials Science and Engineering **621** (2019) 012019 doi:10.1088/1757-899X/621/1/012019

The assessment of Danau Kota Lake water quality using chemometrics approach

Hafizan Juahir[1,2*], Adiana Ghazali[1*], Azimah Ismail[1], Mahadzirah Mohamad[3], Firdaus Mohamad Hamzah[4], Sunardi Sudianto[5], Mohd LokmanMohd Lasim[6], Mohamed Aidi Shahriz[7]

[1]East Coast Environmental Research Institute (ESERI), Universiti Sultan Zainal Abidin (UniSZA), Gong Badak Campus, 21300 Kuala Nerus, Terengganu, MALAYSIA.

[2]Faculty of Bioresources and Food Industry, Universiti Sultan Zainal Abidin (UniSZA), Besut Campus, 22200 Besut,Terengganu,MALAYSIA.

[3]Faculty of Business Management and Accountancy, Universiti Sultan Zainal Abidin (UniSZA), Gong Badak Campus, 21300 Kuala Nerus, Terengganu, MALAYSIA.

[4]Unit of Fundamental Engineering Studies, Faculty of Engineering and Built Environment, UniversitiKebangsaanMalaysia, 43600 UKM Bangi, Selangor, MALAYSIA.

[5]Department of Biology and Institute of Ecology, Faculty of Mathematic and Natural Sciences, UniversitasPadjadjaran, Jalan Raya Bandung-Sumedang Km. 21, Jatinagor, Sumedang 45363, West Java, INDONESIA.

[6]Kuala Lumpur City Hall, Menara DBKL 1, Jalan Raja Laut, 50350 Kuala Lumpur, MALAYSIA.

[7]Impian EkuitiSdn. Bhd., No.20B, Jalan MJ/1, Taman Maju Jaya, JalanKlang Lama, 46000 Petaling Jaya, Selangor, MALAYSIA.

*Corresponding authors E-mail: hafizanjuahir@unisza.edu.my, adiana.ghazali@gmail.com

Abstract. Danau Kota Lake is public attraction area, located in the middle area of Kuala Lumpur. An excessive count of visitors visiting the lake has weakening the lake water quality. The present study was driven to evaluate the lake water quality for environmental conservation. Some water quality and metals data were in-situ measured whereas some were collected for laboratory analysis. BOD, OG, AN and TC are identified as major pollutants towards Danau Kota Lake water quality. The decreasing concentration of DO and increasing of turbidity towards the deeper lake water column has been deteriorated the living water communities and caused hypoxic condition in the bottom water of Danau Kota Lake. Spatial variation of metals in the lake sediment has evidenced the accumulation of metals in the middle part of the lake area due to the sedimentation process. Based on PCA, the possible pollutant sources for this lake are the organic pollutants from the wastewater and industrial pollutants as well as the abundance of suspended algae. Since the water quality of Danau Kota Lake has marked some issues, the present study may assist Kuala Lumpur City Council to prevail the lake water quality challenge promptly by cost saving and in efficient manner.

1. Introduction

A lake usually knows as slow flowing open water with a land surrounding. It is also a group of water bodies that consist of ponds and impoundments. In other term, lakes are absent or lack of exchange with the ocean as they are situated in the inland. They are original from the geographical processes and terminate resulted from the ponding mechanisms with evaporation changing the hydrological balance. Chapman [1] has reviewed numerous result of this mechanism which produces 11 major different lake types and sub-divided into 76 subtypes. These different types of lakes were sharing their ecological and biogeochemical processes which include biological, physical and chemical properties made the lake a complex ecosystem. Because of this, most organisms are depending on the lake due to output which provides a great 'ecosystem services' to the consumer. Nowadays, compilations of studies or research related to the lake are increasing in order to improve a knowledge regarding this subject. Continues gain of knowledge will help to provide a management framework as arise of a problem and issue in the lake dynamic ecosystem [2].

Water catchment development has resulted from a great impact and major transformations of the watershed at the lake have led to the disappearance of some of the ecological component in the lake. Definition of this activities could be defined in two dimensions which through chemical emission and physical degradation [3]. Chemical emission is the product from the pollutant and is solved by the regulation of the system and technology while the physical degradation made up by the environment modification which is needed new and systematic approach. Data collection reflects the change in the surrounding environment which detecting the spatial and temporal changes [4]. This outcome effect of many environment parameters (biotic and abiotic factor) could provide a complete representative change that happens in the lakes Result from the research by Yang et al. [5] reveals that lake natural environment of a lake is a multivariate complex system and difficult to analyze a large number of measured variables and it is required multivariate approaches and metrics for quality assessment. Therefore, a lot of statistical methods like data classification, interpretation and modelling were made to assessing the lake quality parameter [6-8].

Various parameters that present in the water body were detected by many research models and one example is the water quality models (WQMs). This model is valuable tools which supporting the prediction of quality in the water and addressing the characteristic of the waterscape. Most water quality models are created in the best structure for measuring a complex parameter in the water [9]. However, increasing number of parameter that found made research gaps in the water quality models conceded. There is a certain condition that consider will be added some parameter in the water bodies. For example, the effect of rates of flow and wind characteristic could be the effect the water quality parameter. This supported by Jiang et al. [10], simulating of water quality is difficult which only a few studies have recorded that conduct in the different situation. Besides that, most parametric studies need an accurate and high rationality because of the sensitivity of models itself. This change situation could show a full evaluation of the system and this model was expected to produce a high-resolution result when the data collected were applied [11].

Application of environment dataset in the water quality model is important. Eutrophication is the problem that being assessing by the water quality models over 25 years of the research [2]). This becomes a major concern in developing country which gives larges impact to social-economic in quality context of life and human health [12]. An excessive richness of nutrient in the water bodies creates higher biological oxygen demand (BOD) which is indicated a lower level of dissolved oxygen. Increase eutrophication could stress the water quality. This is because; high concentrations of nutrient (nitrogen and phosphorus) in the lake can portray a good water quality. However, with this kind condition will make incapable of producing any useful result in the water quality monitoring. Research is needed to create various techniques to improve water quality as theeutrophication will limit the effectiveness of the models. One of the technique have found are the mixed between various models to produces the best water quality prediction and being revised in the many previous studies [13-14]. overall, water quality needs to be secure in their advantages and should be considered because most parameters in the lake are inherent temporal and spatial variability [15].

The present study was carried out at Danau Kota Lake, Setapak, Kuala Lumpur with the main purpose of investigating the current water quality status of the lake. Besides that, in any research, data analysis iscompulsory in order to extract as much information and interpretation for a precise decision making based on monitored data. The collected water quality data from Danau Kota Lake water column has been compared to Malaysian National Water Quality Standard (NWQS) guided by Malaysian Department of Environment. There are six (6) classes (I, IIA, IIB, III, IV and V) which describe the descending order of water quality as Class I is the 'best' and Class V is the 'worst' [16]. The current survey has compared the Danau Kota Lake water quality against Class II NWQS (suitable for recreational use body contact) due to the fact that this lake is actively dominated by recreational activities. Additionally, the present study was intentionally applied chemometrics techniques on the collected water quality of Danau Kota Lake and the concentration of heavy metals in the lake sediment.

2. Methodology

2.1. Study area

Danau Kota Lake is a recreational lake which is located in the center city of Kuala Lumpur, Malaysia (Figure 1). This area is considerably as one of the densely populated area and tourist attraction area in Peninsular Malaysia. Additionally, the lake surround area is dominated by the domestic, commercial and industrial land-based activities [17-20]. Danau Kota Lake is a well-known recreational lake as it provides approximately 1.2 km of track length circling around the lake for walking, jogging and cycling. Since this lake is surrounded with various human activities, the lake water is facing a critical pollution issue as the lake sometimes produce unpleasant smell and some irresponsible visitors has been littering around the area which is inconvenient situation for other visitors.

Figure 1. Water quality sampling point at Danau Kota Lake.

2.2. Sample and in-situ data collection

The present study has selected 10 sampling location within Danau Kota Lake area (Figure 1). Samples and in-situ data were collected spatially and temporally on the 27th and 30th of December 2016 as well as on the 3rd to 4th of January 2017. 15 water quality parameters were collected from Danau Kota Lake water during day and night time at all stations. Water quality parameters namely pH, temperature, dissolved oxygen (DO), total dissolved solid (TDS), ammoniacal nitrogen (AN), total nitrogen (TN) and total phosphate (TP) were in-situ measured using YSI ProMultimeter. In the meantime, turbidity, biological oxygen demand (BOD) and chlorophyll-a (Chl-a) were also in-situ measured, using HACH 2100 Portable Turbidity Meter, Modernwater Portable BOD Meter and Hydrolab DS 5x Sonde respectively. Besides that, some water samples were collected for laboratory analysis in order to define the level of chemical oxygen demand (COD), total suspended solid (TSS), oil and grease (OG), Escherichia coli (E-coli) and total coliform (TC). Analysis for the listed laboratory test water quality parameters were carried out using HACH Method 8000 (COD Reactor

Model HACH DRB 200) for COD, gravimetric method (APHA 2540-D) for TSS, USEPA Method 1664b for OG as well as USEPA Method 1604 for E-coli and TC. Meanwhile, the concentration of 22 metals were defined from the lake sediment of five (5) sampling locations. Those parameters are arsenic (As), mercury (Hg), aluminium (Al), cadmium (Cd), chromium (Cr), copper (Cu), iron (Fe), lead (Pb), magnesium (Mg), manganese (Mn), selenium (Se), Argentum (Ar), tin (Sn), zinc (Zn), barium (Ba), boron (B), calcium (Ca), nickel (Ni), potassium (K), strontium (Sr), vanadium (V) and cobalt (Co). Danau Kota Lake sediment was digested using mixed concentrated acids and analysed using Inductively Coupled Plasma Mass Spectrometry in order to define the concentration of heavy metals [21-24].

2.3. Statistical analysis

The data matrix in this analysis involves [11 (parameters) x 10 (sampling points)] for surface water quality analysis; [5 (parameters) x 6 (different depths)] for different depth of water quality analysis; as well as [22 (parameters) x 5 (sampling points/sediment samples)] for lake sediment analysis.

2.4. Cluster analysis (CA)

In this study, Hierarchical Aligned Cluster Analysis (HACA) was employed to investigate the grouping of the sampling sites (spatial). HACA is a common method to classify [25] variables or cases (observations/samples) into clusters with high homogeneity level within the class and high heterogeneity level between classes with respect to a predetermined selection criterion [26-27]. Ward's method using Euclidean distances as a measure of similarity [28-30] within HACA has proved to be a very efficient method. The result is illustrated by a dendogram, presenting the clusters and their proximity [31]. The Euclidean distance (linkage distance) is reported as *D*link / *D*max, which represents the quotient between the linkage distance divided by the maximal distance. The quotient is usually multiplied by 100 as a way to standardize the linkage distance represented by the *y*-axis [32].

2.5. Discriminant analysis (DA)

Discriminant analysis determines the variables that discriminate between two or more naturally occurring groups/clusters. It constructs a discriminant function (DF) for each group [33-34]. DFs are calculated using equation (1):

$$f(G_i) = k_i + \sum_{j=1}^{n} w_{ij} P_{ij} \ , \tag{1}$$

where *i*is the number of groups (*G*), *ki*is the constant inherent to each group, *n* is the number of parameters used to classify a set of data into a given group, and *wj*is the weight coefficient assigned by DF analysis (DFA) to a given parameter (p_j).

In this study, DA was applied to determinewhether the groups differ with regard to the mean of a variable and to use that variable to predict group membership. Two groups for temporal analysis (two different sampling time are day and night), which standard, forward stepwise and backward stepwise modes. These were used to construct DFs to evaluate temporal variations in the lake water quality. The sampling time (temporal) were the grouping (dependent) variables, while all the measured parameters constitute the independent variables. In the forward stepwise mode, variables are included step by step beginning with the most significant variable until no significant changes were obtained. In the backward stepwise mode, variables are removed step by step beginning with the less significant variable until no significant changes were obtained.

2.6. Principal Component Analysis (PCA)

The most powerful pattern recognition technique that is usually coupled with HACA is the PCA. It provides information on the most significant parameters due to spatial and temporal variations that describes the whole data set by excluding the less significant parameters with minimum loss of original information [32]. The principle component (PC) can be expressed as:

$$z_{ij} = a_{i1}x_{1j} + a_{i2}x_{2j} + ... + a_{im}x_{mj} , \tag{2}$$

where z is the component score, a is the component loading, x is the measured value of the variable, i is the component number, j is the sample number, and m is the total number of variables.

The FA is usually applied as a method to interpret a large complex data matrix and offers a powerful means of detecting similarities among variables or samples [35]. The PCs generated by PCA are sometimes not readily interpreted; therefore, it is advisable to rotate the PCs by varimax rotation. Varimax rotations applied on the PCs with eigenvalues more than 1 are considered significant [36-37] in order to obtain new groups of variables called varimax factors (VFs). The number of VFs obtained by varimax rotations is equal to the number of variables in accordance with common features and can include unobservable, hypothetical, are considered as "strong"; 0.75–0.50, as "moderate"; and 0.50–0.30, as "weak" significant factor loadings [38].

Source identification of different pollutants was made on the basis of different activities in the surrounding lake area in light of previous literatures. The basic concept of Factors Analysis (FA) is expressed as:

$$z_{ij} = a_{i1}f_{1j} + a_{i2}f_{2j} + ... + a_{im}f_{mj} + e_{ij}, \tag{3}$$

where z is the measured value of a variable, a is the factor loading, f is the factor score, e is the residual term accounting for errors or other sources of variation, i is the sample number, j is the variable number, and m is the total number of factors.

In this study, PCA and FA was applied to the data sets of the lake water quality laboratory analysis (11 variables) for the 10 different sampling stations (points). The temporal variation of lake water quality was carried out due to two different sampling time which is day and night. The input data matrices (variables × cases) for PCA and FA were 11x10 for both sampling periods, day and night.

3. Results and discussion

3.1. Danau Kota Lake water quality criteria assessment based on Class IIB Malaysian NWQS

Danau Kota Lake water quality which has been observed spatially at 10 sampling stations were compared to Class IIB NWQS. Table 1 shows that lake water quality data which is comply (below limit) and not-comply (above limit) Class IIB NWQS for both sampling period i.e. during the day and night time.Based on the comparison, it is clearly demonstrating that, BOD, OG, AN and TC are 100% not comply with NWQS during daytime monitoring of Danau Kota Lake water quality samples. DO concentration is not complying with Class IIB NWQS during the day time samples at nine (9) stations except for station SP1. All DO, BOD, OG and AN data collected during the night time period are not complying with Class IIB NWQS. Besides that, 56% (61 count data) and 48% (53 count data) of temperature, pH, TSS and TP are complying with Class IIB NWQS during day and night time period respectively.

IOP Conf. Series: Materials Science and Engineering **621** (2019) 012019 doi:10.1088/1757-899X/621/1/012019

Table 1. The assessment of observed Danau Kota Lake water quality in comparison with Class IIB NWQS.

Station	DO	pH	BOD	COD	TSS	OG	AN	TN	TP	Chl-a	TC	E.coli
DAY												
SP1	below	below	above	below	below	above	above	above	below	above	above	above
SP2	above	below	above	above	below	above	above	above	below	below	above	below
SP3	above	below	above	below	below	above	above	below	below	above	above	below
SP4	above	below	above	above	below	above	above	above	below	above	above	below
SP5	above	below	above	above	below	above	above	above	below	below	above	above
SP6	above	below	above	above	below	above	above	below	below	below	above	above
SP7	above	below	above	below	below	above	above	below	below	below	above	above
SP8	above	below	above	above	below	above	above	above	below	below	above	above
SP9	above	below	above	below	below	above	above	above	below	below	above	above
SP10	above	below	above	below	below	above	above	below	below	below	above	above
NIGHT												
SP1	above	below	above	above	below	above	above	above	below	below	below	below
SP2	above	above	above	above	below	above	above	below	below	below	above	below
SP3	above	below	above	below	below	above	above	below	below	below	above	below
SP4	above	below	above	below	below	above	above	below	below	below	above	below
SP5	above	below	above	above	below	above	above	below	below	below	above	below
SP6	above	below	above	below	below	above	above	below	below	below	above	above
SP7	above	below	above	below	below	above	above	below	below	above	above	below
SP8	above	below	above	above	below	above	above	above	below	above	above	above
SP9	above	below	above	above	below	above	above	above	below	below	above	below
SP10	above	below	above	above	below	above	above	below	below	below	above	below
Limit	5-7	6.0-9.0	3	25	50	0.04	0.3	5	5	10	50000	5000

Note: below limit; above limit

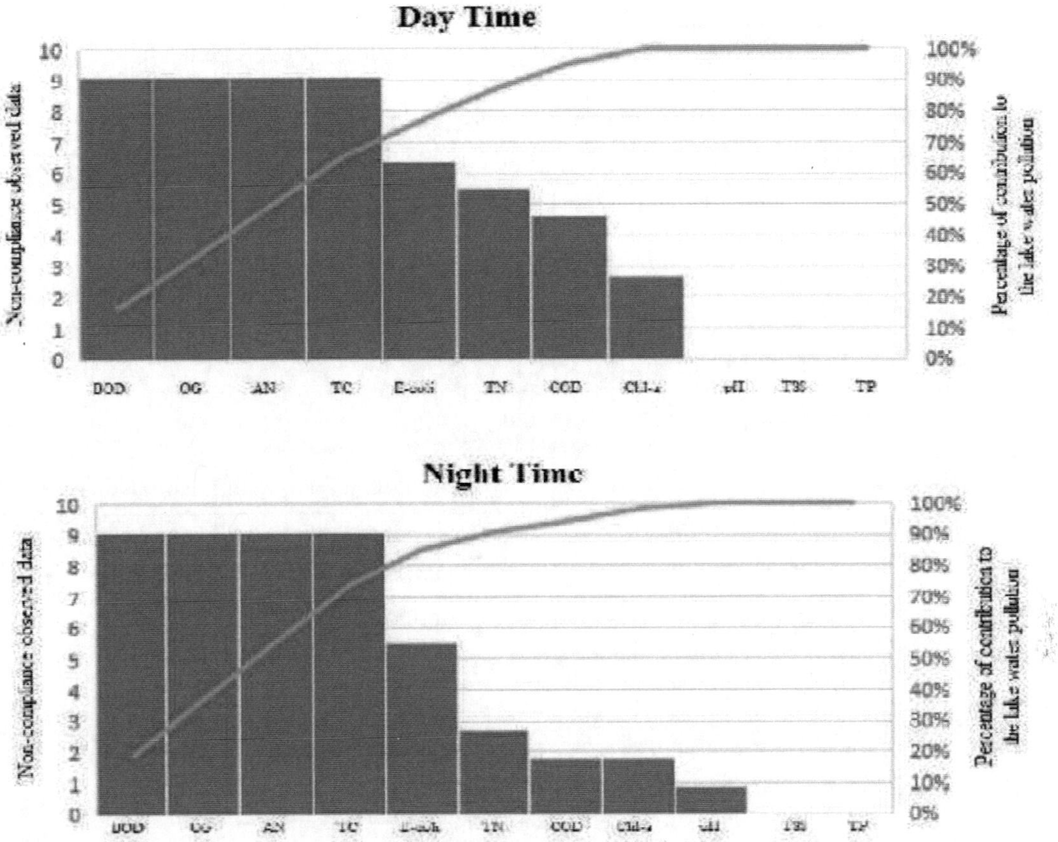

Figure 2. Pareto charts of major contributors toward Danau Kota Lake water pollution during both sampling periods.

Pareto charts in Figure 2 highlight the major pollutants which contribute to the serious water problems in Danau Kota Lake. Those pollutants are BOD, OG, AN and TC with high cumulative percentage recorded of 66% and 74% both sampling periods. Therefore, the present study has drawn the need of local authority to pay particular attention in controlling these pollutants, particularly on the untreated direct discharge into the Danau Kota Lake water system from the surrounding area.

3.2. Spatial classification of Danau Kota Lake water quality

The 15 water quality parameters observed at 10 sampling stations in Danau Kota Lake were classified spatially using HACA. Dendograms in Figure 3 give two clusters during day time period and three clusters during night time period. Based on day time dendogram, Cluster 1 is consisting of eight sampling points namely SP1, SP2, SP3, SP4, SP5, SP7, SP9, and SP10. In the meantime, Cluster 2 is consisting of two sampling points which are SP6 and SP8. For the time being, based on the night sampling period, Cluster 1 consists of eight sampling points which are SP1, SP3, SP4, SP6, SP7, SP8, SP9 and SP10. Besides that, Cluster 2 and Cluster 3 consist of one sampling point which is SP2 and SP5 respectively. Therefore, in consideration for future sampling plan, it is sufficient to collect samples from only two (2) sampling points which represent each cluster during day time period as well as three (3) sampling points which represent three (3) clusters during the night time period.

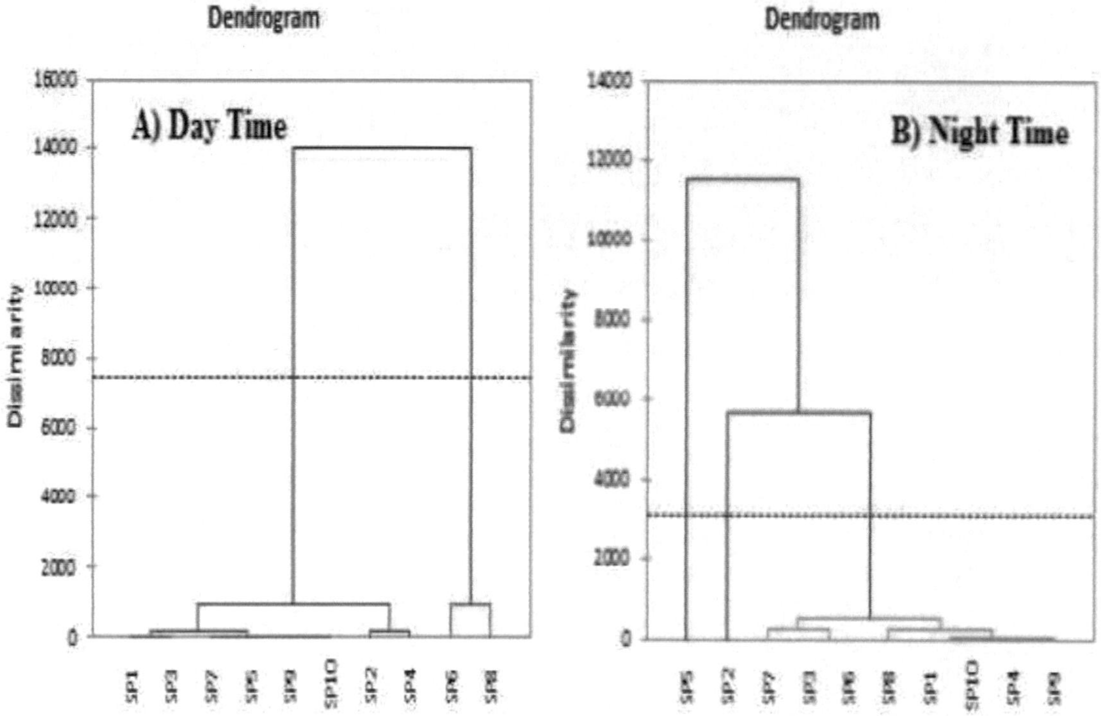

Figure 3. Spatial classification of water quality during different sampling periods.

3.3. Spatial classification of Danau Kota Lake sediment based on metal variations.

The spatial classification of Danau Kota Lake sediment was performed based on the concentration of 22 metals. However, only 17 metals were taken into consideration for further classification analysis, as other five (5) were removed due to no significant variation throughout the sampling period. Dendrogram in Figure 4 shows that there are three (3) clusters were formed spatially based on the similarity of metals concentration in Danau Kota Lake sediment. The clusters can be interpreted as High Contaminated Sampling Point (HCSP) which consists of station SP2 and SP4; Moderate Contaminated Sampling Point (MCSP) which comprises of station SP5; as well as Low Contaminated Sampling Point (LCSP) which involves of station SP3 and SP10 (Figure 5). Based on the spatial cluster analysis on metals concentration in the sediment sample; in order to monitor the future metals contamination in Danau Kota Lake sediment, it is strongly suggested to collect the sediment samples at three (3) stations which represents three (3) formed clusters. As a consequence, by practicing this method, the future sampling cost and time can be reduced. Other than that, the spatial distribution of metals in Danau Kota Lake sediment has emphasized the polluted area was within the middle part of the lake area. Thus it is stronglypresuming that metals input from the surrounding area was accumulated during the lake sedimentation process towards the middle part of the lake area.

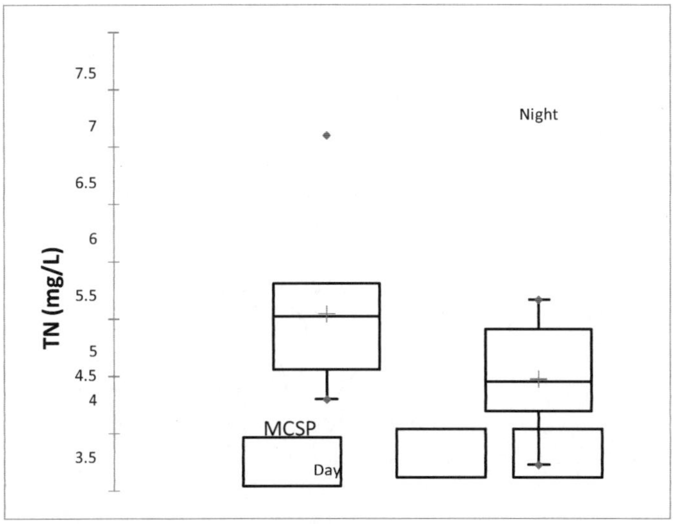

Figure 4. Spatial classification of metals variation in Danau Kota Lake sediment using HACA.

3.4. *Spatial variation of water quality based on different water depth*

Table 2 and Figure 6 show the spatial distribution of lake water quality according to the water depth. Based on the observed data, pH and DO were enriched at the surface water but depleted at the bottom water. On the other hand, the concentration of TDS was low at the surface water but high at the bottom water of Danau Kota Lake. In the meantime, temperature and turbidity did not show any particular trend. Additionally, data analysis has shown the lowest concentration of pH and DO as well as an increased of TDS was observed at the deepest depth of Danau Kota Lake water column. Obviously, the difference of pH levels recorded was corresponded to different water levels or depth of Danau Kota Lake as itcould be due to the increased of carbon dioxide (CO_2) level from the respiratory activities and the decomposition of organic materials below the thermocline. Hence the saturated concentration of CO_2 accumulates in the lower strata of the lake. However, the concentration of DO increases as the pressure increases. DO saturation decreases by 10% per meter as depth of lake increases, resulting from the hydrostatic pressure. To these, the fact is that, the concentration of dissolved oxygen decreases downward to the lake bottom [35,38].

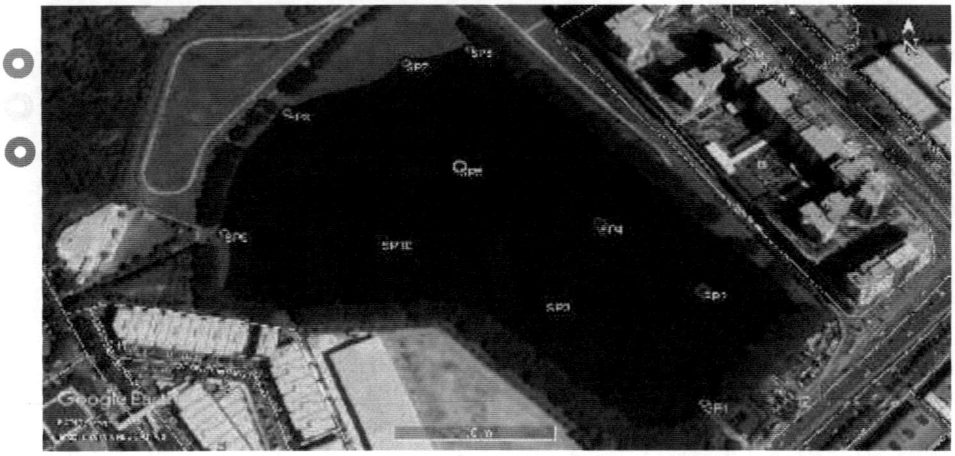

Figure 5: Spatial classification of metals distribution in Danau Kota Lake sediment.

IOP Conf. Series: Materials Science and Engineering **621** (2019) 012019 doi:10.1088/1757-899X/621/1/012019

In the common notion of water quality, the high levels of TDS lead to the increasing of water temperature and decreasing of DO concentration.It is commonly attributed to the absorbance of more heat from solar radiation by the suspended particles [35].The absorbed heat is then directly transferred to the surrounding water through the conduction process. The warmer water unlikely has potential to hold as much as DO compare to cold water, resulting in a significant decline in DO levels. In addition, as the surface temperature increases, it can cause the stratification, or layering of lake-water. Furthermore, whenever thelake water stratifies, the upper and lower layers will form different layers. No mixture between the layers.The frequent occurrence of decomposition and respiration in the lower layers of lake- water creates the condition of hypoxic (low level of DO). Therefore, it will aggravate the living communities or organisms in the lake to survive.

Based on data analysis, temperature and turbidity are not statistically significant (p>0.05). Temperature tends to deplete in the deeper water depth, in contrary with turbidity where it was spatially increased from the water surface downward to the lake bottom.

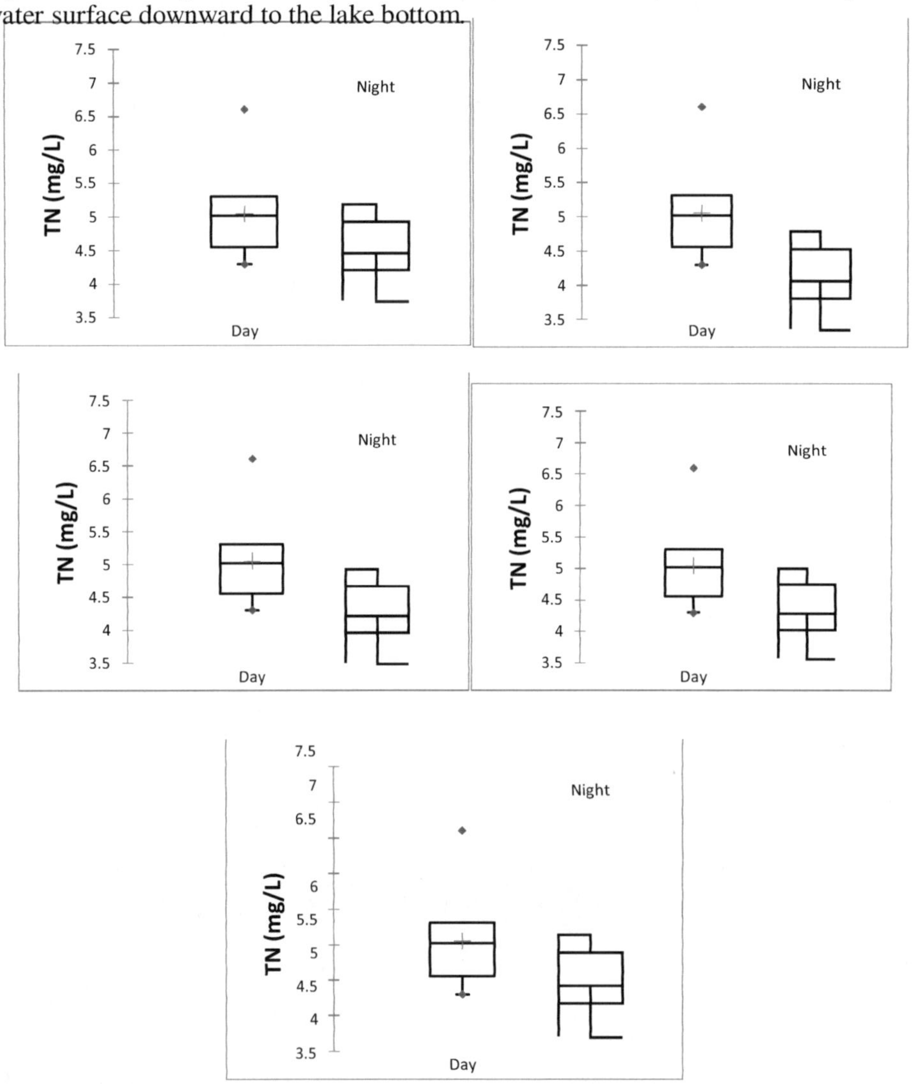

Figure 6. Spatial water quality of Danau Kota Lake at different water depth.

Table 2. Mann-Kendall non parametric test of trend on Danau Kota Lake water quality based on different water depth.

Variable	Kendall's tau	S	Var(S)	Trend	p-value
pH	-0.905	-19.000	0.000	↓	0.003
Temp.	-0.524	-11.000	0.000	NT	0.136
DO	-0.905	-19.000	0.000	↓	0.003
Turbidity	0.238	5.000	0.000	NT	0.562
TDS	0.714	15.000	0.000	↑	0.030

Note: ↑-upward trend; ↓-downward trend; NT-no trend

3.5. Temporal Variation of Danau Kota Lake Water Quality

3.5.1. The variation of water quality parameters temporally using DA.

In order to study the temporal variation of water quality data, DA was applied to the raw data post grouping of Danau Kota Lake water quality. The collected water quality data was grouped into two main clusters/groups, defined by day and night time sampling period. Both groups were treated as the dependent variables, while the water quality parameters were treated as the independent variables. DA wasperformed through modes of standard, forward stepwise, and backward stepwise modes with calculated DFA of 94.4% (11 discriminant variables), 72.2% (one discriminant variable), and 94.4% (three discriminant variables), respectively (Figure 8 and Table 3). The forward stepwise mode has shown that only pH found to be the significant variable. This indicates that pH have high variation in terms of their temporal distribution. pH shows a lower value during the day time (μ_{Day} = 7.12) but slightly higher during the night time (μ_{Night} = 8.13). On the other hand, the backward stepwise mode of which included TSS (μ_{Day} = 20 mg/L; μ_{Night} = 7 mg/L) and TN (μ_{Day} = 5.05 mg/L; μ_{Night} = 4.48 mg/L) as the second and third parameter show a high temporal variation. Both parameters show slightly higher concentrations during day time period compared to night time period. This pattern was anticipated due to the watering activities of landscape plants and ornaments home during daytime.

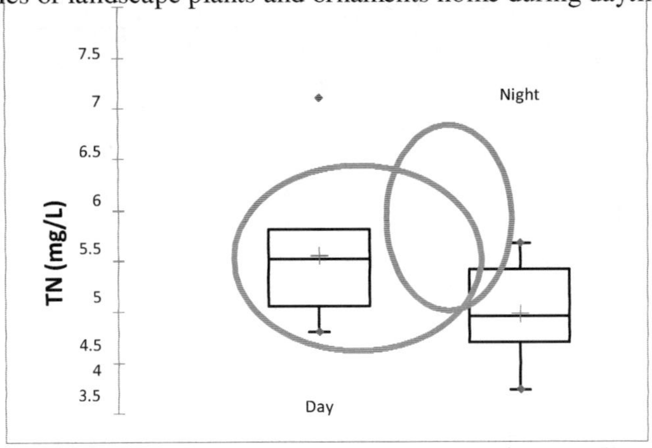

Figure 7. Plot of discriminant functions for temporal variation of Danau Kota Lake water quality.

Apart from that, the residual of chemical fertilizer with high contents of nitrogen and plant soil transported by water residual during the watering activities has been flushed into the drain and finally directly goes into the lake systems. In the meantime, the remaining eight water quality parameters depicted no significant difference ($p>0.05$) between both sampling periods. Box and whisker plots of pH, TSS and TN recorded during the sampling period are shown in Figure 8. These parameters which yielded the highest variability (the most significant ($p<0.05$)) was analysed using backward stepwise of the DA and the output will be used for further discussion.

Table 3. Classification matrix of temporal DA for Danau Kota Lake water quality.

Time	Time Assign by DA		Total	% correct
	Day	Night		
Standard mode DA				
Day	9	0	9	100.00%
Night	1	8	9	88.89%
Total	10	8	18	94.44%
Stepwise forward DA				
Day	7	2	9	77.78%
Night	3	6	9	66.67%
Total	10	8	18	72.22%
Stepwise backward DA				
Day	9	0	9	100.00%
Night	1	8	9	88.89%
Total	10	8	18	94.44%

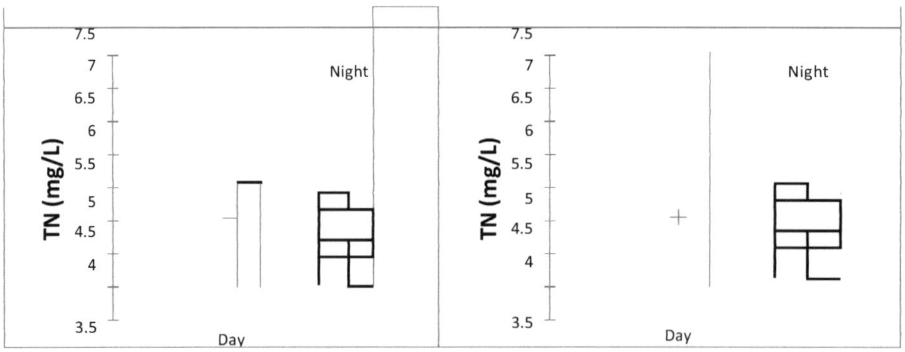

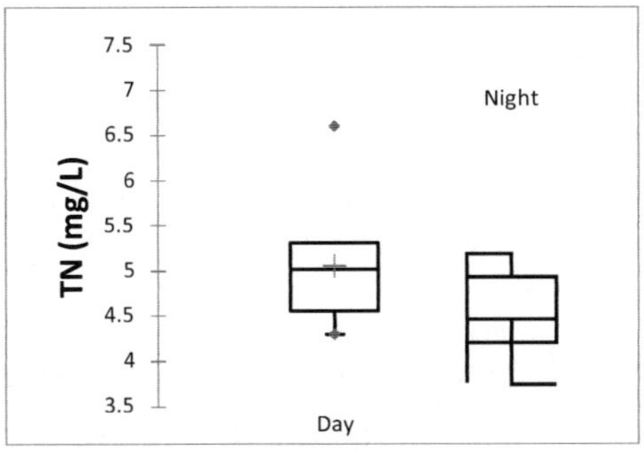

Figure 8. Box and whisker plots of significant parameters of temporal variation for Danau Kota Lake waters.

3.5.2. Identification of the possible pollution sources using PCA

PCA was employed to the data set to compare the compositional patterns between the examined water parameters and to identify the factors that influence the variation of lake water quality. Three PCs were obtained with eigenvalues larger than one summing almost 87.6% of the total variance in the data set. Corresponding varimax factors (VFs), variable loadings, and variance explained are presented in Table 4.

Table 4. Loadings of varimax-rotated PCs for water quality data collected in Danau Kota Lake.

Variables	VF1	VF2	VF3
pH	-0.379	0.306	-0.127
BOD	0.978	-0.011	-0.176
COD	0.973	0.059	0.024
TSS	0.707	-0.535	-0.418
O&G	0.98	-0.008	-0.138
AN	-0.627	0.197	0.558
TN	0.205	0.955	-0.056
TP	0.976	-0.123	-0.058
Ch-a	-0.116	-0.076	0.917
Coliform	0.964	0.171	-0.195
E.coli	0.91	0.299	-0.252
Eigenvalue	7.056	1.481	1.102
Variability (%)	64.147	13.464	10.014

Cumulative %	64.147	77.612	87.626

For daytime, VF1 is accounted for 64.2% of the total variance, reveals the strong positive loadings on the variables of BOD (0.98), COD (0.97), TSS (0.71), O&G (0.98), TP (0.98), coliform (0.95) and E.coli (0.91). Strong positive loadings on BOD and COD prove the influence of organic pollutants from the point sources. These pollutants are commonly related to the anthropogenic pollution sources such as wastewater or sewage treatment plants, domestic wastewater and industrial effluents [39-41].

TSS can be originated from both wastewater treatment plants (point source, PS) and polluted non-point source (NPS) [42-43].Solids are also known to be one of the most common contaminants found in urban storm water. Commonly they are originated from many sources which include the erosion of pervious surfaces and dust; litter and other particles deposited on the impervious surfaces from anthropogenic activities and atmospheric deposition.Additionally, the soil erosion at construction sites and robust development is counted as a he major source of TSS into the lake water column.Therefore, the urban waterways create more TSS which lead to the increase of turbidity and reduce the penetration of light in the lake water column. Consecutively, low penetration of light in the water cause limitation in the growth of desirable aquatic plants.

Other than that, TSS which settling inside the lake ecosystem as the bottom deposits will lead to the sedimentation, alter and eventually destroys the habitat of many fish species and bottom-dwelling organisms [39-40,44]. Besides that, the settling down of TSS on the lake bed surface provides a medium for the accumulation, transport and storage of other pollutants including nutrients, pesticides, metals as well as OG pollutants. According to Bastian [45], the typical concentration of TSS in urban runoff is substantially higher than that in treated wastewater. Therefore, the construction produces the highest loading of TSS over other urban land use categories. In the meantime, OG pollutant sources are most possibly derive from roads, driveways, parking lots, vehicle maintenance areas, gas stations, illicit dumping into storm drains and domestics cooking wastes [46].

TP is the main nutrients found in urban storm water. The major sources of nutrients in urban storm water are lawn fertilizers, atmospheric deposition, automobile exhaust, soil erosion, animal waste and detergents [47]. TP measures the total amount of phosphorus in both the organic and inorganic forms. The degree to which nitrogen and phosphorus are present in a lake can determine the trophic status and the amount of algal biomass produced. Excess nutrients tend to increase the primary biological productivity. The major impact associated with nutrient over enrichment is excessive growth of algae that leads to nuisance algal blooms and eutrophic conditions [48-49]. A secondary impact is the residual negative effect of decomposing algae in the form of sediment oxygen demand that depletes dissolved oxygen concentrations, particularly in bottom waters [50].

TC and E. coli which enters the lake water system are most possibly related to municipal sewage and wastewater treatment plants as well as domestic wastewater [51]. The presence of coliform and E. coli in lake water consumes a large amount of oxygen. As the amount of available DO decreases, they will undergo anaerobic fermentation processes which leading to the production of ammonia and organic acids [52]. The hydrolysis of these acidic materials causes a decrease of water pH values [53].

VF2 explains 13.5% of the total variance with a strong loading on TN. TN pollution is a crucial to the entire Klang Valley region. The main caused are identified from the untreated excreta in sewage directly discharged into the lake. In addition, the TN presence in the environment from the decomposition of the organic matter, such as food residue discarded directly into the waterways [54].

VF3 explains 10% of the total variance with 0.917 strong loading on Chl-a. The high loading of the Chl-a could be derived from the abundance of suspended algae within the lake. The accumulation of the Chl-a in the water can reduce the dissolve oxygen that harmful to the various species of living organism [50].

4. Conclusion and recommendation

4.1. Conclusion

The spatial and temporal study of Danau Kota Lake is important in order to enhance the water quality of the lake it is located in the middle of Kuala Lumpur city center. The present study is important study as it could aid the Kuala Lumpur City Council to come up with a pragmatic and precise decision making, conserve and preserveDanau Kota Lake water column. CA has classified the polluted area in Danau Kota Lake into three (3) groups namely HCSP, MCSP and LCSP. In the meantime, the concentration of DO decreased downward to the lake bottom, which is aggravating the living communities in the lake. It could be due to the decomposition and respiration in the lower layers of lake-water. Therefore, it creates the condition of hypoxic (low dissolved oxygen levels). The output of temporal DA backward stepwise mode reveals that three variables namely pH, TSS and TN has influenced the lake water quality during the night time period. Through PCA, strong positive loadings were shown by BOD (0.98), COD (0.97), TSS (0.71), O&G (0.98), TP (0.98), coliform (0.95) and E.coli (0.91). the strong positive loadings on these variables could possibly derived from the anthropogenic pollution sources such as wastewater or sewage treatment plants, domestic wastewater and industrial effluents. Since Danau Kota Lake is a public attraction lake, it is important to maintain the physical view of the lake water. High level of turbidity lead to a murkier lake water which is unpleasant view for the visitors. Additionally, high turbidity also cause limitation in sunlight penetration into the water column which impact the growth of desirable aquatic plants. Therefore, the present study could help Kuala Lumpur City Council to overcome the lake water quality problem with expeditiously, cost saving and efficient manners.

4.2. Recommendation

The Kuala Lumpur City Council should implement the water quality monitoring program to Danau Kota Lake and other lakes within the Klang Valley in the agenda of future plan towards achieving the environment sustainability. By maintaining the water quality of the lake, the lake itself safe for any form of outdoor water activities and aesthetically preserved. There are few recommendations should be taken into Kuala Lumpur City Council consideration based on the present study findings, as follow: (i) On-ground verification as to reconfirm the point source of contamination flowing into the lake for the next course of treatment. The flow has to be treated before flowing into the lake; (ii) Control the point-sources prior the flow being treated.

To find the best water treatment technology to solve the water quality problem efficiently in order to meet the Class IIB NWQS.

5. Acknowledgement

Kuala Lumpur City Hall (DBKL) funded this research through a research grant (grant numbers: I1017-0008). The authors would like to thank DBKL for the research grant and the East Coast Environmental Research Institute (ESERI) for the assistance given during sampling and analysis.

References

[1] Chapman D V 1996 Water quality assessments: a guide to use of biota, sediments and water in environmental monitoring.

[2] Brönmark C and Hansson L A 2002 Environmental issues in lakes and ponds: current state and perspectives. *Environmental conservation*, 29(3), 290-307.

[3] Zalewski M 2007 Ecohydrology as a concept and management tool. *Managing Water in Diverse Ecosystems to Ensure Human Well-being. United Nations University-International Network on Water, Environment and Health (UNU-INWEH)*, Hamilton, pp39-50.

[4] Matthews G B, Matthews R A and Hachmöller B 1991 Mathematical analysis of temporal and spatial trends in the benthic macroinvertebrate communities of a small stream. *Canadian Journal of Fisheries and Aquatic Sciences*, 48(11), 2184-2190.

[5] Yang G, Ma R, Zhang L, Jiang J, Yao S, Zhang M and Zeng H D 2010 Lake status, major problems and protection strategy in China. *Journal of Lake Sciences*, 22(6), 799-810.

[6] Simeonov V, Stratis J A, Samara C, Zachariadis G, Voutsa D, Anthemidis A and Kouimtzis T 2003 Assessment of the surface water quality in Northern Greece. *Water research*, 37(17), 4119-4124.

[7] Massart D L, Vandeginste B G M, Buydens L M C, De Jong S, Lewi P J and Smeyers-Verbeke J 1998 *Handbook of chemometrics and qualimetrics; data handling in science and technology, parts A and B.*, Elsevier: Amsterdam.

[8] Zhou F, Liu Y and Guo H 2007 Application of multivariate statistical methods to water quality assessment of thewatercourses in Northwestern New Territories, Hong Kong. *Environmental Monitoring and Assessment*, 132(1-3), 1-13.

[9] Wang Y, Wang P, Bai Y, Tian Z, Li J, Shao X and Li BL 2013 Assessment of surface water quality via multivariate statistical techniques: a case study of the Songhua River Harbin region, China. *Journal of Hydro-Environment Research*, 7(1), 30-40.

[10] Jiang L, Li Y, Zhao X, Tillotson M R, Wang W, Zhang S, Sarpong L, Asmaa Q and Pan B 2018 Parameter uncertainty and sensitivity analysis of water quality model in Lake Taihu, China. *Ecological Modelling,* 375, 1-12.

[11] Hughes D A and Slaughter A R 2016 Disaggregating the components of a monthly water resources system model to daily values for use with a water quality model. *Environmental Modelling & Software*, 80, 122-131.

[12] Tundisi J G, Santos M A and Menezes C F S 2003 Tucuruí reservoir and hydroelectric power plant.Sharing Experiences and Lessons Learned in Lake Basin Mangement, Burlington, *Vermont. Management Experiences and Lessons Learned Brief*, 1, 1-20.

[13] Luo Y, Yang K, Yu Z, Chen J, Xu Y, Zhou X and Yang Y 2017 Dynamic monitoring and prediction of Dianchi Lake cyanobacteria outbreaks in the context of rapid urbanization. *Environmental Science and Pollution Research*, 24(6), 5335-5348.

[14] Wei Z, Han Y, Li M, Yang K, Yang Y, Luo Y and Ong SH 2017 A small UAV based multi-temporal image registration for dynamic agricultural terrace monitoring. *Remote Sensing,* 9(9), 904.

[15] Missaghi S, Hondzo M and Melching C 2013 Three-dimensional lake water quality modeling: sensitivity and uncertainty analyses. *Journal of environmental quality*, 42(6), 1684-1698.

[16] Ismail A, Toriman ME, Juahir H, Zain SM, Habir NLA, Retnam A and Azid A 2016 Spatial assessment and source identification of heavy metals pollution in surface water using several chemometric techniques. *Marine pollution bulletin*, 106(1), 292-300.

[17] Mohamad M, Ali A M, Ab Ghani N I, Abdullah A R and Mokhlis S 2012 Positioning Malaysia as a tourist destination based on destination loyalty. *Asian Social Science* 9(1), 286-292.

[18] Awang Z, Afthanorhan A, Mohamad M and Asri M A M 2015 An evaluation of measurement model for medical tourism research: The confirmatory factor analysis approach. *International Journal of Tourism Policy* 6(1), 29-45.

[19] Mohamad M, Ghani N I A, Mamat M and Mamat I 2014 Satisfaction as a mediator to the relationships between destination image and loyalty. *World Applied Sciences Journal* 30(9), 1113-1123.

[20] Kashif M, Samsi S Z M, Awang Z and Mohamad M 2016 EXQ: measurement of health care experience quality of Malaysian settings: A contextualist perspective. *International Journal of Pharmaceutical and Healthcare Marketing* 10(1), 27-47.

[21] Ong M C, Joseph B, Shazili N A M, GhazaliA and Mohamad M N 2015 Heavy Metals Concentration in Surficial Sediments of Bidong Island, South China Sea off the East Coast of Peninsular Malaysia. *Asian Journal of Earth Sciences*, 8(3), 74

[22] Ghazali A, Shazili N A M, Bidai J and Shaari H 2016 The spatial distribution of Al, Fe, Cu, Cd and Pb in the surface sediment of Brunei Bay, Borneo during the southwest and northeast monsoons. *Journal of Sustainability Science and Management*, 11, 93-106

[23] Bidai J, Adiana G and Shazili N A M 2016 Particle sizes effect of the elements in the South China Sea sediment off Pahang coastal during the Northeast monsoon (pre-) and the Southwest monsoon (post-) periods. *Environmental Earth Sciences*, 75(7), 614

[24] Adiana G, Shazili N A M and Marinah M M 2011 Cadmium, manganese and lead distribution in the South China Sea off the South Terengganu Coast, Malaysia during post-monsoon and pre-monsoon. *Journal of Sustainability Science and Management*, 6(2), 181-192

[25] Massart D L and Kaufman L 1983 The interpretation of analytical data by the use of cluster analysis. NewYork:Wiley.

[26] McKenna J E Jr 2003 An enhanced cluster analysis program with bootstrap significance testing for ecological community analysis. *Environmental Modelling& Software,* 18(2), 205–220.

[27] Adiana G, Juahir H, Joseph B and Shazili N A M 2017 Tracing the sources of lead (Pb) in Brunei Bay, Borneo by using integrated spectrometry ICP-MS and chemometric techniques. *Marine pollution bulletin*, 123(1-2), 232-240

[28] Willet P 1987 *Similarity and clustering in chemical information systems*. New York: Research Studies Press,Wiley.

[29] Adams M J 1998 The principles of multivariate data analysis. *In*Ashurst PR & Dennis MJ (Eds), *Analytical methods of food authentication,* Blackie Academic & Professional, London, pp350.

[30] Otto M 1998 Multivariate methods. *In* R. Kellner, J.M.Mermet, M. Otto, & H. M. Widmer (Eds.), *Analyticalchemistry*. Wenheim:Wiley-VCH.

[31] Forina M, Armanino C and Raggio V 2002 Clustering with dendograms on interpretation variables. *AnalyticaChimicaActa,* 454, 13–19.

[32] Singh K P, Malik A and Sinha S 2005 Water quality assessment and apportionment of pollution sources of Gomti River (India) using multivariate statistical techniques: A case study. *AnalyticaChimicaActa, 35*, 3581–3592.

[33] Johnson R A and Wichern D W 1992 *Applied multivariate statistical analysis* (3rd ed.). Prentice-Hall Int.:New Jersey.

[34] Juahir H, Sharifuddin M Z, MohdKamil Y, TengkuHanidza T I, MohaArmi A S, Toriman M E & Mokhtar M 2011 Spatial water quality assessment of Langat River Basin (Malaysia) using environmetric techniques. *Environmental Monitoring Assessment* 173, 625-641.

[35] Reghunath R, Murthy S T R and Raghavan B R 2002 The utility of multivariate statistical techniques in hydrogeochemical studies: An example from Karnataka, India. *Water Research, 36*, 2437–2442.

[36] Adiana G, Joseph B, Shazili N A M, Mohamad M and Juahir H 2018 Spectrometry and chemometricsinterpretation of metals in the Sulu and Celebes Seas of Malaysian waters during PMSE 09' Expedition. *Journal of Engineering and Applied Sciences* 13(7), 1745-1756.

[37] Kim J-O & Mueller C W 1987 *Introduction to factor analysis: What it is and how to do it. Quantitative applications in the social sciences series*. Newbury Park: Sage University Press.

[38] Liu C W, Lin K H and Kuo Y M 2003 Application of factor analysis in the assessment of groundwater quality in a Blackfoot disease area in Taiwan. *The Science of the Total Environment,* 313, 77–89.

[39] Al-Odaini N A, Zakaria M P, Zali M A, Juahir H, Yaziz M I and Surif S 2012 Application of chemometrics in understanding the spatial distribution of human pharmaceuticals in surface water. *Environmental Monitoring and Assessment* 184(11), 6735-6748.

[40] Juahir H, Zain S M, Aris A Z, Yusof M K, Samah M A A and Mokhtar M 2010 Hydrological trend analysis due to land used changes at Langat River Basin. *Environment Asia* 3, 20-31.

[41] Mohd K A K, Mohd E T and Nur H S 2015 Classification of tropical river using chemometrics technique: Case study in Pahang River, Malaysia. *Malaysian Journal of Analytical Sciences* 19(5), 1001-1018.

[42] Dolan D M 1993 Point source loadings of phosphorus to Lake Erie: 1986-1990. *Journal of Great Lakes Research* 19(2), 212-223.

[43] Ha S R and Bae M S 2001 Effects of land use and municipal wastewater treatment changes on stream water quality. *Environmental Monitoring and Assessment* 70(1-2), 135-151.

[44] Kamarudin M K A, Toriman M E, Rosli M H, Juahir H, Aziz N A A, Azid A and Sulaiman W N A 2015 Analysis of meander evolution studies on effect from land use and climate change at the upstream reach of the Pahang River, Malaysia. *Mitigation and Adaptation Strategies for Global Change* 20(8), 1319-1334.

[45] Bastian R K 1997 Biosolids management in the United States. *Water Environment & Technology* 9(5), 45-50.

[46] Rozell D J and Reaven S J 2011 Water pollution risk associated with natural gas extraction from the Marcellus Shale. *Risk Analysis* 32(8), 1382-1393.

[47] Dodds W K & Smith V H 2016 Nitrogen, phosphorous and euthropication in streams. *International Society of Limnology* 6, 155-164.

[48] Dodds W K, Bouska W W, Eitzmann J L, Pilger T J, Pitts K L, Riley A J, Schloesser J T and Thombrugh D J 2009 Eutrophication of U.S. freshwaters: Analysis of potential economic damages. *Environmental Science and Technology* 43(1):12-19.

[49] Smith V H, Tilman G D and Nekola J C 1999 Eutrophication: impacts of excess nutrient inputs onfreshwater, marine and terrestrial ecosystems.*Environmental Pollution*100(1-3), 179-196.

[50] Withers P J A and Jarvie H P 2008 Delivery and cycling of phosphorous in rivers: A review. *Science of the Total Environment* 400(1-3), 379-395.

[51] Frenzel S A and Couvillion C S 2002 Fecal-indicator bacteria in streams along a gradient of residential development. *Journal of the American Water Resources Association* 38(1), 265-273.

[52] Mara D 2005 Pond process design - a practical guide. *In* Andy Shilton (Ed). *Pond Treatment Technology* IWA Publishing, UK, pp100-131.

[53] Alrumman S A, El-kott A F and Keshk S M A S 2016 Water pollution: Source and treatment. *American Journal of Environmental Engineering* 6(3), 88-98.

IOP Conf. Series: Materials Science and Engineering **621** (2019) 012019 doi:10.1088/1757-899X/621/1/012019

[54] Hamilton A J, Stagnitti F, Xiong X, Kreidl S L, Benke K K and Maher P 2007 Wastewater irrigation: The state of play. *Vadose Zone Journal* 6(4), 823-840.

IOP Conf. Series: Materials Science and Engineering **621** (2019) 012020 doi:10.1088/1757-899X/621/1/012020

Modeling and analysis of CO2 emissions in million tons of sectoral greenhouse gases in Indonesia

Ruly Budiono[1], Hafizan Juahir[2], Mustafa Mamat[3], Sukono[4*], Sudradjat Supian[4], Muhamad Nurzaman[1]

[1]Department of Biology, Faculty of Mathematics and Natural Sciences, Universitas Padjadjaran, Indonesia

[2]East Coast Environmental Research Institute (ESERI), Universiti Sultan Zainal Abidin, Malaysia

[3]Faculty of Informatics and Computing, Universiti Sultan Zainal Abidin, Malaysia

[4]Department of Mathematics, Faculty of Mathematics and Natural Sciences, Universitas Padjadjaran, Indonesia

E-mail: ruly@unpad.ac.id; hafizanjuahir@unisza.edu.my; musmat567@gmail.com; sukono@unpad.ac.id: sudradjat@unpad.ac.id; m.nurzaman@unpad.ac.id

*Corresponding author: sukono@unpad.ac.id

Abstract. Economic sector development is an effort to increase national economic growth which is usually measured by increasing Gross Domestic Product (GDP). But economic sector development affects CO2 emissions in million tons of sectoral greenhouse gases. Therefore, this paper performs modeling and analysis of CO2 emissions in million tons of sectoral greenhouse gases in Indonesia. Sectors analysed include: industrial, commercial, household, transportation, power plant, and others. The objective is to obtain a model estimator and determine the level of sectoral CO2 emissions contribution to total CO2 emissions in Indonesia. Modeling is done by using Cobb-Douglas multiplicative production function approach. While to make estimation of model parameters is done by using Ordinary Least Square (OLS) method. The analysis results show that sectors: industrial, commercial, household, transportation, power plant, significantly contribute to total CO2 emissions. While the other sectors do not significantly affect the total CO2 emissions. The analysis results also show that the five-sector CO2 emission contribution model for total CO2 emissions significantly follows Cobb-Douglas multiplicative function, with correlation strength level which is 99.9% coefficient of determination. Thus, the model estimators obtained are excellent and can be used to determine the level of sectoral CO2 emissions contribution to total CO2 emissions in Indonesia.

1. Introduction

Economic sector is a framework of development of national monetary growth. Successful development of economic growth is usually measured by raising Gross Domestic Product (GDP). National economic development is a process of raising total and per capita income, which is accompanied by fundamental changes in the economic structure of a country and the equitable distribution of income for its people [13]. Through economical sector development, it is possible to change the agrarian economic structure into industrial economic structure. So that the economic activities undertaken by the state will be more diverse and dynamic [14; 15]. However, unplanned economic sector development may impact damage to the environment. Economic development through the industrial sector has indeed brought prosperity to its people, but has also given a negative impact on the ecological system of the world [3; 17]. It is now felt that the world community is in a great crisis situation. The world community has felt its own bad consequences of air pollution and global warming caused by the build-up of CO2 emissions in the air [4; 10; 18]. CO_2 emissions have the most important role in climate change because these gases continue to accumulate in the atmosphere in large numbers [5; 9; 22]. Therefore, it is emerging to study the effect of sectoral economic growth on total CO_2 emission increase.

Filiz and Omer [7], has undertaken research on long-term and causal relationships between the industrial, cement and steel production sectors; power plants; oil consumption; and increased CO_2 emissions in Turkey. The study was conducted using an auto-regression vector testing approach (VAR) in the period 1990 to 2010. The results showed that there are two way linkage of Granger causality between CO_2 emissions and cement production and power plant. While the results of impulse-response, indicating that CO_2 emissions largely influenced by the industrial production sector, followed by cement production sectors, power plants, oil consumption, and steel production. Aye and Edoja [2] examined the effect of economic sector growth on CO_2 emission changes using panel data analysis consisting of 31 developing countries. The analysis results show that the growth of the economic sector has a negative effect on changes in CO_2 emissions under low growth conditions, but has a positive effect on high growth conditions, with a marginal effect higher in high growth conditions. Coi and Abdullah [6], predicted changes in CO_2 emissions using linear regression. CO_2 emission change trends are related to the development of economic sectors and other variables, such as demand and supply sectors in the economy and energy consumption sectors. They say that the linear model is one of the common methods used to explain the correlation between CO_2 emissions and related economic sector variables. Similar studies have also been conducted by Khobai & Roux [11], Mrabet et al. [16], and Shukla [19]. Kosztowniak [12] analysed the influence of foreign direct investment net entry on GDP in Poland between 1994-2012, using the Cobb-Douglas production function. In such an analysis defined conditions necessary for the positive influence of foreign direct investment in Poland. Also assumed the assumptions are the Cobb-Douglas production function and predicted changes in GDP value in Poland. In the analysis also identified factors that significantly affect economic growth in Poland. A similar study has also been conducted by Anghelache et al. [1], Gogtay et al. [8], Shukla [19], and Wang & Fu [21].

Referring to the above description, this paper intends to model and analyse CO_2 emissions in million tons of sectoral greenhouse gases in Indonesia. The economic sectors analysed include: industrial, commercial, household, transportation, power plant, and others. The purpose of this study is to obtain a model estimator and determine the level of CO_2 emissions contribution from economic sector development to the total CO_2 emissions in Indonesia. The estimated model is expected to be used in the determination of CO_2 emission level control policy from sectoral economic development activities to CO_2 emissions.

2. Methodology

The research methodology here is a scientific process or way of getting the facts by a step or rule that can use to get an estimator of the correlation model between sectoral CO_2 emissions with total CO_2 emissions. After several observations, it is suspected that the correlation of sectoral CO_2 emissions to total CO_2 emissions follows the Cobb-Douglas production function model. Therefore, this section discusses the methods including: Cobb-Douglas production function, parameter estimation method, goodness of fit test, and forecasting.

2.1. Cobb-Douglas production function

The form of production function is the interaction between the input (input) and the output (output). There are several forms of production functions such as the Cobb-Douglas production function. Referring to Wang & Fu [21] and Kosztowniak [12], generally, the Cobb-Douglas is a multiplicative production function with more than two independent variables can be given as follows:

$$G = \phi Z_1^{\beta_1} Z_2^{\beta_2} Z_3^{\beta_3} \ldots Z_k^{\beta_k} e^{\varepsilon}, \tag{1}$$

where G is the dependent variable (output); ϕ is coefficient intercept; Z_1, Z_2, Z_3, ..., Z_k are independent variables (input); $\beta_1, \beta_2, \beta_3, ..., \beta_k$ are coefficient of elasticity of independent variables; $e = 2.7182818285$ natural numbers; and ε is error (residual).

The sum of the elasticity coefficients can be used as a measurement of returns to scale. There are 3 characteristics of size return to scale, as follows:

- Decreasing returns to scale, when $\sum_{i=1}^{k} \beta_i < 1$
  Indicates that the Cobb-Douglas production function has additional characteristics of diminishing returns on a production scale, where output increases with a proportion smaller than the proportion of inputs.
- Constant returns to scale, when $\sum_{i=1}^{k} \beta_i = 1$
  Indicates that the Cobb-Douglas production function has a constant yield characteristic on the scale of production, when all inputs rise in certain proportions, then the produced output rises in exact proportion to the proportion of the input.
- Increasing returns to scale, when $\sum_{i=1}^{k} \beta_i > 1$
  Indicates that the Cobb-Douglas production function has additional characteristics of increased yield on a production scale, where output increases with a proportion greater than the proportion of inputs.

When the left and right sides of equation (1) are transformed by natural logarithms, then equation (2) is obtained, and resulting the following linear regression equation:

$$\ln G = \ln \phi + \beta_1 \ln Z_1 + \beta_2 \ln Z_2 + \beta_3 \ln Z_3 + ... + \beta_k \ln Z_k + \varepsilon \quad (2)$$

Furthermore, if we give $Y = \ln G$, $\beta_0 = \ln b$, $X_1 = \ln Z_1$, $X_2 = \ln Z_2$, $X_3 = \ln Z_3$, ..., $X_k = \ln Z_k$, then equation (2) is a multiple linear regression equation which can be expressed as follows [1]:

$$Y = \beta_0 + \beta_1 X_1 + \beta_2 X_2 + \beta_3 X_3 + ... + \beta_k X_k + \varepsilon . \quad (3)$$

The estimation of the multiple linear regression equation (3) is expressed as follows:

$$\hat{Y} = \beta_0 + \beta_1 X_1 + \beta_2 X_2 + \beta_3 X_3 + ... + \beta_k X_k . \quad (4)$$

So that, the residual values can be determined as $\varepsilon = Y - \hat{Y}$.

2.2. *Method of parameter estimation*

In this section we discuss the method of parameter estimation from multiple regression equation in general. Once the estimation is done by using the matrix equation approach, the multiple linear regression equation (3) can be given as follows [1]:

$$\mathbf{Y} = \mathbf{X}\boldsymbol{\beta} + \mathbf{e} . \quad (5)$$

Where:

$$\mathbf{Y} = \begin{bmatrix} Y_1 \\ Y_2 \\ \vdots \\ Y_n \end{bmatrix}; \mathbf{X} = \begin{bmatrix} 1 & X_{12} & X_{13} & \cdots & X_{1k} \\ 1 & X_{22} & X_{23} & \cdots & X_{2k} \\ \vdots & \vdots & \vdots & \ddots & \vdots \\ 1 & X_{n2} & X_{n3} & \cdots & X_{nk} \end{bmatrix}; \boldsymbol{\beta} = \begin{bmatrix} \beta_0 \\ \beta_1 \\ \vdots \\ \beta_k \end{bmatrix}; \boldsymbol{\varepsilon} = \begin{bmatrix} \varepsilon_1 \\ \varepsilon_2 \\ \vdots \\ \varepsilon_n \end{bmatrix},$$

where $\mathbf{Y}$ is matrix of $(n \times 1)$, $\mathbf{X}$ is matrix of $(n \times (k+1))$, $\boldsymbol{\beta}$ is matrix of $((k+1) \times 1)$, and $\boldsymbol{\varepsilon}$ is matrix of $(n \times 1)$.

Based on Anghelache et al. [1] and Gogtay et al. [8], to obtain the estimator values of matrix parameters, it can be determined by using ordinary least square method (OLS), i.e. by minimizing the residual squares amount which the equation is as follows:

$$\text{Minimization } \sum \varepsilon_i^2 = \boldsymbol{\varepsilon}^T \boldsymbol{\varepsilon} = (\mathbf{Y} - \mathbf{X}\boldsymbol{\beta})^T (\mathbf{Y} - \mathbf{X}\boldsymbol{\beta}) . \tag{6}$$

Where $\boldsymbol{\varepsilon}^T = (\mathbf{Y} - \mathbf{X}\boldsymbol{\beta})^T$ is matrix transpose of $\boldsymbol{\varepsilon}$. Because $\boldsymbol{\beta}^T \mathbf{X}^T \mathbf{Y}$ is a scalar, so its value is equal to its transpose, i.e. $\mathbf{Y}^T \mathbf{XB}$. In the process of minimizing, the equation (6) is obtained as follows:

$$\frac{\partial \sum \varepsilon_i^2}{\partial \boldsymbol{\beta}} = -2\mathbf{X}^T \mathbf{Y} + 2\mathbf{X}^T \mathbf{X}\boldsymbol{\beta} = 0 . \tag{7}$$

Further, solving equation (7) can be obtained the parameter estimator matrix as follows:

$$\boldsymbol{\beta} = (\mathbf{X}^T \mathbf{X})^{-1} \mathbf{X}^T \mathbf{Y} . \tag{8}$$

Where the matrix $(\mathbf{X}^T \mathbf{X})^{-1}$ is the inverse of the matrix $(\mathbf{X}^T \mathbf{X})$.

Note that this approach method can only be used if the matrix $(\mathbf{X}^T \mathbf{X})$ has an inverse.

2.3. *Test of goodness of fit*

After the parameter estimator values of multiple linear regression models are obtained, the parameter estimator need to be tested its goodness of fit at a defined level of significance. The goodness of fit test is to determine that the model is significantly able to describe the actual data. Goodness of fit test to model parameter estimator is done by using partial significance test, simultaneous significance test, assumption of residual normality test, and coefficient of determination test.

- Test of individual significance parameters

The individual parameter estimator significance test is performed to test each parameter β_i ($i = 0,1,2,..., k$) of equation (8), individually in affecting the dependent variable. To test parameter estimators β_i, determine hypothesis H_0: $\beta_i = 0$ and alternate hypotheses H_1: $\beta_i \neq 0$. Testing is done by using statistic t_{stat}. the equation is as follows:

$$t_{Stat} = \frac{\beta_i}{SE(\beta_i)}, \tag{9}$$

where $SE(\beta_i)$ is the standard error of parameters β_i ($i = 0,1,2,..., k$).

The test criterion is to reject the hypothesis H_0 and accept the hypothesis H_1, if statistic $|t_{Stat}| > |t_{(n-k,\frac{1}{2}\alpha)}|$ or statistic $\Pr[t_{Stat}] < \alpha$, where statistic $t_{(n-k,\frac{1}{2}\alpha)}$ is the critical value of the distribution of t at the level of significance $100(1-\alpha)\%$, k is the number of parameters, and n is the number of data. Instead, accept the hypothesis H_0 and reject the hypothesis H_1, if statistic $|t_{Stat}| \leq |t_{(n-k,\frac{1}{2}\alpha)}|$ or statistic $\Pr[t_{Stat}] \geq \alpha$, where statistic $t_{(n-k,\frac{1}{2}\alpha)}$ is the critical value of the distribution of t at the level of significance $100(1-\alpha)\%$, k is the number of parameters, and n is the number of data [1; 20].

- Test of the significance of parameters simultaneously

Test the significance of parameter estimators simultaneously, is conducted to simultaneously test parameter estimators β_i ($i = 0,1,2,..., k$) of equation (4), in affecting the dependent variable. In this test is determined the hypothesis $H_0: \beta_0 = \beta_1 = \beta_2 = ... = \beta_k = 0$ and alternate hypotheses $H_1: \exists\ \beta_0 \neq \beta_1 \neq \beta_2 \neq ... \neq \beta_k \neq 0$. Testing is done by using statistic F, the equations is as follows:

$$F_{stat} = \frac{\text{MS}_{\text{Reg}}}{\text{MS}_{\text{Error}}}, \tag{10}$$

where MS_{Reg} mean square due to regression, and MS_{Error} mean square due to residual variation.

The test criterion is to reject the hypothesis H_0 and accept the hypothesis H_1, if statistic $F_{Stat} > F_{(k-1,n-k,1-\alpha)}$ or statistic $\Pr[F_{Stat}] < \alpha$, where statistic $F_{(k-1,n-k,1-\alpha)}$ is the critical value of the distribution of F at the level of significance $100(1-\alpha)\%$, k is number of parameters, and n is the number of data. Instead, accept the hypothesis H_0 and reject the hypothesis H_1, if statistic $F_{Stat} \leq F_{(k-1,n-k,1-\alpha)}$ or statistic $\Pr[F_{Stat}] \geq \alpha$, where statistic $F_{(k-1,n-k,1-\alpha)}$ is the critical value of the distribution of F at the level of significance $100(1-\alpha)\%$, k is number of parameters, and n is the number of data [8; 20].

- Test of assumption of residual normality

The normality assumption test is performed to determine that the residual data is distributed following a normal distribution. Normality assumption test can be done using Kolmogorov-Smirnov (KS) statistic. In this test, the hypothesis is determined H_0: normally distributed data, and alternatives of hypotheses H_1: data is not normally distributed. Testing is done by determining residual standard deviation by using equation as follows:

$$S_{\varepsilon_i} = \sqrt{\frac{\sum_{i=1}^{n}(\varepsilon_i - \bar{\varepsilon})^2}{n-1}}. \tag{11}$$

Then, transformed residual value ε_i become z_i using equation $z_i = (\varepsilon_i - \bar{\varepsilon}) / S\varepsilon_i$. The probability value $P(z_i)$ is determined using standard normal distribution tables. While the probability $S(z_i)$ is determined using the equation $S(z_i) = randl(z_i)/n$. Next, we calculate the values of absolute difference $|S(z_i) - P(z_i)|$. The statistical value of Kolmogorov-Smirnov KS_{Stat} is determined using the following formula [20]:

$$KS_{Stat} = \max\{|S(z_i) - P(z_i)|\}. \tag{12}$$

Next, the statistical value KS_{Stat} is compared with statistical critical value of $KS_{(\alpha,n-1)}$, at the level of significance $\alpha = 0.05$. The test criterion is to reject the hypothesis H_0 and accept the hypothesis H_1, if statistic $KS_{Stat} > KS_{(\alpha,n-1)}$, where n is the number of data. Instead, accept the hypothesis H_0 and reject the hypothesis H_1, if statistic $KS_{Stat} \leq KS_{(\alpha,n-1)}$, where n is the number of data [20].

- Coefficient of determination

Referring to Anghelache et al. [1] and Gogtay et al. [8], the coefficient of determination R^2 is used to measure how much the variability of independent variables to the dependent variable, based on the level of correlation power. So the coefficient of determination value is the ability or strength of the independent variable X_i $(i = 1,2,...,k)$ in affecting the dependent variable Y. Value of the coefficient of determination R^2 is determined using the following equation:

$$R^2 = \frac{\sum_{i=1}^{n}(\hat{Y}_i - \bar{Y})^2}{\sum_{i=1}^{n}(Y_i - \bar{Y})^2}. \tag{13}$$

The value of determination coefficient is ranged between 0 and 1. The value of determination is small close to 0 which means the variation of independent variable to the dependent variable is very weak. Conversely, a value close to 1 means that the variation of the independent variable to the dependent variable is very strong.

2.4. *Forecasting (Prediction)*

Forecasting (prediction) is very important in modeling. Forecasting is done where the complexity and uncertainty are faced by the model maker. So as to require the level of error as small as possible or the level of accuracy as big as possible, there are many methods that can be used to measure the error rate of a forecasting model. One of which is Mean Absolute Percentage Error (MAPE). Determining the MAPE value is determined by using the equation as follows:

$$MAPE = \left(\frac{1}{n} \sum_{i=1}^{n} \frac{|Y_i - \hat{Y}_i|}{Y_i} \right) \times 100\% \tag{14}$$

The smaller the MAPE value, the smaller the value of error, which also means the greater the accuracy of the model in predicting an event [20].

3. Result and Analysis

In this section we will discuss the result and analysis which include: data analysed, multiple linear regression model estimation, and estimation of Cobb-Douglas model, and forecasting (prediction).

3.1. *Data analysed*

In this study the data analysed are CO2 emissions in million tons of sectoral greenhouse gases in Indonesia. It is a secondary data obtained from the Ministry of Energy and Mineral Resources of the Republic of Indonesia, in the period 2000 to 2015. Data are distributed from CO2 emissions emitted by each sector of the economy which includes: industrial, commercial, household, transportation, and power generation. The data can be shown as the graph given in Figure 1.

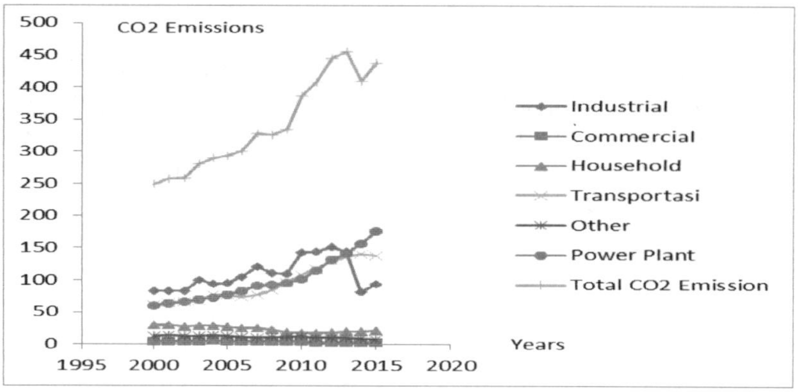

Figure 1. Graph of CO2 emissions of each sector and total

Taking into account to Figure 1, it appears that the total CO2 emissions emitted by the economic sector, from year to year, it tends to increase with a high trend. The contribution of total CO2 emissions from the largest economic sector is from the industry. The next contribution is from power generation, transportation, household, commercial and others. The contribution of each economic sector fluctuates with a relatively constant trend. Furthermore, based on the data in this study, it carried out modeling the contribution of CO2 emissions from each sectoral to total CO2 emissions from year to year, using Cobb-Douglas model. While for parameter estimation, it is done by using multiple linear regression equation, ordinary least square method (OLS).

3.2. Multiple linear regression modeling

In this section we estimate the multiple linear regression models of natural logarithm data from CO2 emissions emitted by sectors: industrial, commercial, household, transportation, power generation, and others, to the natural logarithm of total CO2 emissions. Estimation is done by using Ordinary Least Square (OLS) method approach which refers to equation (8). The estimation of multiple linear regression models is done by using Minitab 16 software, and the result of estimation process as given in Table 1.

Table 1. Results of estimation of multiple linear regression model of CO2 emission

```
The regression equation is
ln(Total CO2 Emission) = 1.49 + 0.282 ln(Industrial) - 0.180 ln(Commercial)
                         + 0.127 ln(Household) + 0.319 ln(Transportation)
                         + 0.110 ln(Other) + 0.251 ln(Power Plant)

Predictor               Coef   SE Coef      T      P
Constant             1.49100   0.30250   4.93  0.001
ln(Industrial)       0.28228   0.01507  18.74  0.000
ln(Commercial)      -0.17962   0.09575  -1.88  0.093
ln(Household)        0.12674   0.02685   4.72  0.001
ln(Transportasi)     0.31914   0.07033   4.54  0.001
ln(Other)            0.10970   0.12420   0.88  0.400
ln(Power Plant)      0.25113   0.08329   3.02  0.015

S = 0.00629268   R-Sq = 99.9%   R-Sq(adj) = 99.9%

Analysis of Variance

Source          DF       SS       MS        F      P
Regression       6  0.65494  0.10916  2756.65  0.000
Residual Error   9  0.00036  0.00004
Total           15  0.65530
```

By Noting the results of the estimation process in Table 1, If the significance level of the level is equal to $\alpha =$ 0.05, then the estimator parameters: Constant, ln(Industrial), ln(Household), ln(Transportation), and ln(Power Plant), have a probability $P <$ 0.05. Unless the estimator of the parameters ln(Commercial) and ln(Other) has a probability $P \geq 0.05$, it means estimators of the parameters ln(Commercial) and ln(Other) individually significant have no effect on the dependent variable ln (Total CO2 Emissions). Therefore, gradually the non-significant parameter estimators are excluded from the multiple linier regression model estimators. We begin with the parameter estimator having the greatest probability, that is, the estimator of the coefficient parameter ln(Other), and then the re-estimation process is done. The re-estimation process is also performed using Minitab 16 software, and the estimation results are given in Table 2.

Table 2. Result of re-estimation process without independent variable ln(Other)

```
The regression equation is
ln(Total CO2 Emission) = 1.71 + 0.286 ln(Industrial) - 0.105 ln(Commercial)
                        + 0.118 ln(Household) + 0.374 ln(Transportation)
                        + 0.187 ln(Power Plant)

Predictor              Coef   SE Coef      T      P
Constant            1.71190   0.16810  10.18  0.000
ln(Industrial)      0.28587   0.01435  19.93  0.000
ln(Commercial)     -0.10505   0.04459  -2.36  0.040
ln(Household)       0.11783   0.02460   4.79  0.001
ln(Transportasi)    0.37434   0.03184  11.76  0.000
ln(Power Plant)     0.18695   0.04018   4.65  0.001

S = 0.00622290   R-Sq = 99.9%   R-Sq(adj) = 99.9%

Analysis of Variance

Source          DF       SS       MS        F      P
Regression       5  0.65491  0.13098  3382.42  0.000
Residual Error  10  0.00039  0.00004
Total           15  0.65530
```

Based on the results of the re-estimation given in Table 2, it appears that the probability of the parameter estimator of all values is less than 0.05. This indicates that there is no longer an estimate to be issued, and thus no further re-estimation is required. Based on the values in Table 2, and with reference to equation (3) multiple linear regression models the estimation results can be expressed in terms of equations as follows:

$$Y = 1.7119 + 0.28587X_1 - 0.10505X_2 + 0.11783X_3 + 0.37434X_4 + 0.18695X_5 + \varepsilon \,. \qquad (15)$$

Where Y is ln(total CO2 emissions), X_1 is ln(Industrial), X_2 is ln(Commercial), X_3 is ln(Household), X_4 is ln(Transportation), X_5 is ln(Power Plant), and ε is residual.

Furthermore, equation (15) needs to be tested its goodness of fit. *First*, the individual significance test is performed on the parameter estimators $\hat{\beta}_0 = 1.71190$, $\hat{\beta}_1 = 0.28587$, $\hat{\beta}_2 = -0.10505$, $\hat{\beta}_3 = 0.11783$, $\hat{\beta}_4 = 0.37434$, and $\hat{\beta}_5 = 0.18695$. The test is performed by referring to equation (9), using a significance level $\alpha = 0.05$. At the level of significance $\alpha = 0.05$ and with degrees of freedom $df = 16\text{-}6 = 10$, from the table of distribution of t, we obtained the critical value of statistic $t_{(10;0.025)} = 2.228$. Based on the estimation results presented in Table 2, the value of statistics is $t_{Stat} = 10.18$ so it is clear that statistic $t_{Stat} > t_{(10;0.025)}$. This shows that parameter estimator $\hat{\beta}_0$ is significant. The significance test also needs to be done for the parameter estimators $\hat{\beta}_1$, $\hat{\beta}_2$, $\hat{\beta}_3$, $\hat{\beta}_4$, and $\hat{\beta}_5$. Testing is also done in the same way, and each result also significantly affects the variable Y (ln(total CO2 emissions)). *Second*, the simultaneous significance test of the parameter estimator $\hat{\beta}_0 = 1.71190$, $\hat{\beta}_1 = 0.28587$, $\hat{\beta}_2 = -0.10505$, $\hat{\beta}_3 = 0.11783$, $\hat{\beta}_4 = 0.37434$, and $\hat{\beta}_5 = 0.18695$. The test is performed by referring to equation (10), using a significance level $\alpha = 0.05$ and degrees of freedom $df = (6\text{-}1;\ 16\text{-}6)$. At the level of significance $\alpha =$

0.05 and with degrees of freedom df = (5; 10), of the table of distribution of F, then we obtained the critical value of statistic $F_{(5;10;0.05)}$= 4.10. Based on the estimation results presented in Table 2, the value of statistic F_{Stat}= 3382.42. So it is clear that statistic $F_{Stat} > F_{(5;10;0.05)}$. This shows that parameter estimators $\hat{\beta}_0$, $\hat{\beta}_1$, $\hat{\beta}_2$, $\hat{\beta}_3$, $\hat{\beta}_4$, and $\hat{\beta}_5$, simultaneously have significant effect on the variable Y (ln(total CO2 emissions)).

Third, the assumption of residual normality assumption, is intended to determine that residuals are normally distributed. Testing is done by referring to equation (12), also by using the level of significance α = 0.05. Based on the results of testing using Minitab 16 software, it gives residual $\varepsilon \sim N(-0.0125, 3.617604)$. Fourth, the determination coefficient test is intended to control the level of strength of the relationship between independent variables with the dependent variable. Determination of coefficient value of determination is done by referring to equation (13). Based on the test results presented in Table 2, it is found that the value of determination R^2 = 99.9%. This shows that the relationship between the independent variable and the dependent variable is very strong.

After a goodness of fit test and all test results are significant, thereby obtaining an estimator of multiple linear regression models of CO2 emissions given in equation (15) can be expressed as equations:

$$\hat{Y} = 1.7119 + 0.28587X_1 - 0.10505X_2 + 0.11783X_3 + 0.37434X_4 + 0.18695X_5 . \quad (16)$$

Furthermore, the model estimator of multiple linear regression equation (16), with reference to equation (1) can be obtained by estimating Cobb-Douglas production function model as follows:

$$\hat{G} = e^{1.7119} Z_1^{0.28587} Z_2^{-0.10505} Z_3^{0.11783} Z_4^{0.37434} Z_5^{0.18695} , \quad (17)$$

where $\hat{G}$ is the output variable estimator (total CO2 emissions); e = 2.7182818285 is a natural number as a multiplier; Z_1 is a variable of industrial sector; Z_2 is variable of commercial sector; Z_3 is variable of household sector; Z_4 is a variable of transport sector; and Z_5 is a variable of power generation sector.

Furthermore, Cobb-Douglas production function estimator (17) is used for forecasting. *First*, forecasting is done by using in-sample data whose results are shown as forecasting graphs as shown in Figure 3. In Figure 3, it appears that the forecasting graph (prediction) coincides with the actual data (total CO2 emissions), and forecasting using the in-sample data gives a MAPE error rate of 0.00394169 or 0.394169%. This means that the model estimator has an accuracy level for forecasting of 99.605831%.

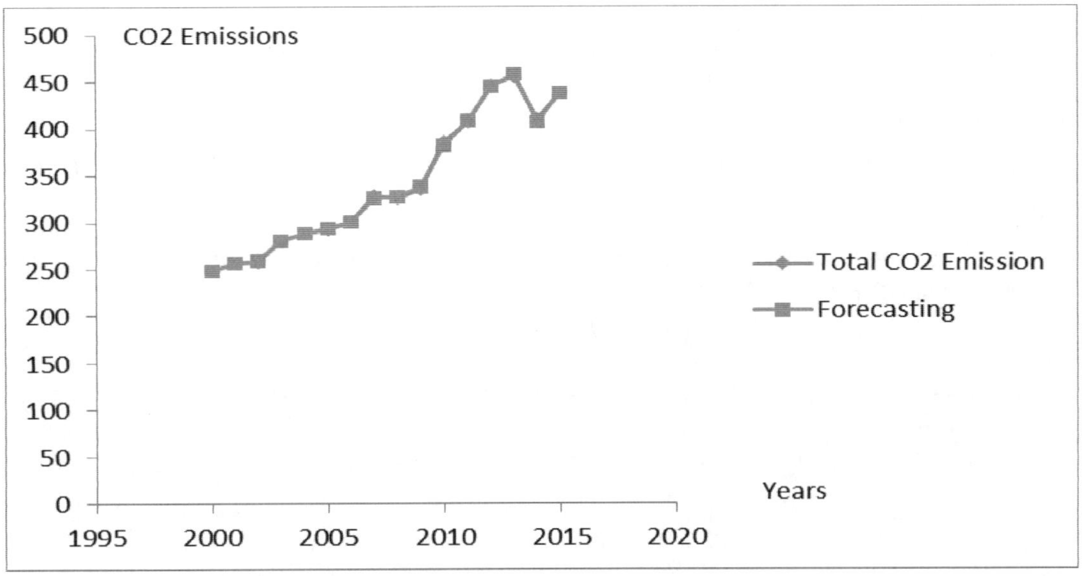

Figure 2. Graph forecasting and actual total CO2 emissions data

Second, forecasting based on out-sample data by estimating the contribution values of sectoral CO2 emissions. For example in certain years, it is estimated that the contribution values of sectoral CO2 emissions as given in Table 3. Using Cobb-Douglas production function estimator (17) obtained the forecast of total CO2 emissions in million tons as given in Table 3.

Table 3. Forecasting of sectoral and total CO2 emissions in million tons in Indonesia

Years	Industrial	Commercial	Household	Transportation	Power Plant	Total CO2 Emissions
2016	105.4	2.13	21.43	140.18	194.44	473.7564790
2017	117.58	1.78	21.85	142.42	213.26	510.9674851
2018	129.76	1.43	22.27	144.66	232.08	550.8038038

4. Conclusion

In this paper we have conducted modeling and analysis of CO2 emissions in million tons of sectoral greenhouse gases in Indonesia. The modeling is done by using Cobb-Douglas multiplicative production function, while for parameter estimation is done by using ordinary least square method (OLS). Based on the analysis results, it can be concluded that sectoral economic growth covering sectors: industrial, commercial, household, transportation, and power plant has resulted in increasing total CO2 emission in Indonesia. The estimated model obtained year by year significantly follows the Cobb-Douglas production function, with a deterministic coefficient of 99.9%. The estimated model of Cobb-Douglas production function model has an insurance rate for forecasting of 99.605831%. Using the estimated Cobb-Douglas production model estimator model, it can be used for forecasting total CO2 emissions in future periods, by estimating the values of sectoral economic variables. If in 2018 it is estimated that the contribution of industrial sector CO2 emission is 129.76; commercial 1.43; household 22.27; transportation 144.66; and 232.08 power plant, the total CO2 emissions will reach 550.8038038 million tons. Therefore, the results of this study are expected to be taken into consideration in environmentally friendly economic development, to support efforts to decrease CO2 emissions in the air.

5. Acknowledgements

Acknowledgments are conveyed to the Rector, Director of Directorate of Research, Community Involvement and Innovation, and the Dean of Faculty of Mathematics and Natural Sciences, Universitas Padjadjaran, with whom the Internal Grant Program of Universitas Padjadjaran was made possible to, fund this research. The grant is a means of enhancing research and publication activities for researchers at Universitas Padjadjaran.

References

[1] Anghelache C, Anghel M G, and Popovic M 2015 Multiple Regressions Used in Analysis of Private Consumption and Public Final Consumption Evolution. *International Journal of Academic Research in Accounting, Finance and Management Sciences*, Vol. 5, No.4, October 2015, pp. 69–73.

[2] Aye G C and Edoja P E 2017 Effect of Economic Growth on CO2 Emission in Developing Countries: Evidence from a Dynamic Panel Threshold Model. *Cogent Economics & Finance*(2017), 5: 1379239, pp. 1-22. https://doi.org/10.1080/23322039.2017.1379239

[3] Asici A A 2011 Economic growth and its impact on environment: A panel data analysis. *MPRA* Paper No. 30238, posted 18. April 2011 12:50 UTC. Online at https://mpra.ub.uni-muenchen.de/30238/.

[4] Budiono R, Juahir H, Mamat M, Sukono and Nurzaman M 2018 Modelling Interaction of CO2 Concentration and the Biomass Algae Due to Reduction of Anthropogenic Carbon Based on Predator-Prey Model. *International Journal of Applied Environmental Sciences*. ISSN 0973-6077 Volume 13, Number 1 (2018), pp. 27-38. © Research India Publications. http://www.ripublication.com

[5] Cioca L I, Ivascu L, Rada E C, Torretta, V., and Ionescu, G. 2015. Sustainable Development and Technological Impact on CO2 Reducing Conditions in Romania. *Sustainability* 2015, *7*, 1637-1650; doi:10.3390/su7021637, pp. 1-14.

[6] Choi C S and Abdullah L 2016 Prediction of Carbon Dioxide Emissions Using Two Linear Regression-based Models: A Comparative Analysis. *Journal of Applied Engineering (JOAE),* 4(4), April-2016 (Volume-IV, Issue-IV), pp. 305-312.

[7] Filiz O and Omer O 2012 An Analysis of CO2 Emissions of Turkish Industries and Energy Sector. *Regional and Sectoral Economic Studies,* Vol.12-2 (2012), pp. 65-85.

[8] Gogtay N J, Deshpande S P and UM Thatte U M 2017 Principles of Regression Analysis. *Journal of The Association of Physicians of India*, Vol. 65, April 2017, pp. 48-52.

[9] Halushchak M, Bun R, Shpak N and Valakh M 2016 Modelling and spatial analysis of greenhouse gas emissions from fuel combustion in the industry sector in Poland. *Econtechmod. An International Quarterly Journal,* 2016. Vol. 5. No. 1, pp. 19–26.

[10] Jin S H, Bai L, Kim J Y, Jeong S J and Kim K S 2017 Analysis of GHG Emission Reduction in South Korea Using a CO2 Transportation Network Optimization Model. *Energies* 2017, 10, 1027, pp. 1-18. doi:10.3390/en10071027.

[11] Khobai H and Roux P L 2017 The Relationship between Energy Consumption, Economic Growth and Carbon Dioxide Emission: The Case of South Africa. *International Journal of Energy Economics and Policy*, 2017, 7(3), pp. 102-109. available at http: www.econjournals.com.

[12] Kosztowniak A 2014 Analysis of the Cobb-Douglas Production Function as a Tool to Investigate the Impact of FDI Net Inflows on Gross Domestic Product Value in Poland in the Period 1994–2012.*Oeconomia Copernicana*, *5*(4), 169-190. https://doi.org/10.12775/ OeC.2014.032.

[13] Knight K W and Schor J B 2014 Economic Growth and Climate Change: A Cross-National Analysis of Territorial and Consumption-Based Carbon Emissions in High-Income Countries. *Sustainability* 2014, *6*, pp. 3722-3731; doi:10.3390/su6063722

[14] Lappalainen S and Kouvo P 2004 Evaluation of Greenhouse Gas Emissions from Landfills in the ST. Petersburg Area-Utilization of Methan in Energy Production, Metgas. *Publication* 10, 2004, pp. 1-56. Lappeenranta University of Technology, Northern Dimension Research Centre. URL: www.lut.fi/nordl.

[15] Mikayilov J I, Marzio G M and Hasanov F J 2018 The Impact Of Economic Growth On Co2 Emissions In Azerbaijan. *Working Paper*, No. 102, February 2018, pp. 1-33. Working Paper Series. ISSN 1973-0381

[16] Mrabet A, Achairi R and Ellouze A 2013 The Two-Way relationship between Economic Growth and CO2 Emissions. *International Conference on Business, Economics, Marketing & Management Research* (BEMM'13) Volume Book: Economics & Strategic Management of Business Process (ESMB), pp. 1-4. Copyright _ IPCO 2013.

[17] Phimphanthavong H 2013 The Impacts of Economic Growth on Environmental Conditions in Laos. *Int.J.Buss.Mgt.Eco.Res.*, Vol 4(5),2013, pp. 766-774. www.ijbmer.com | ISSN: 2229-6247.

[18] Reinecke R and Casey N H 2017 A whole farm model for quantifying total greenhouse gas emissions on South African dairy farms. *South African Journal of Animal Science* 2017, 47 (No. 6), pp. 883-894.

[19] Shukla M 2007 Estimation of CO2 Emissions using Energy Input-Output (EIO) Tables for India. *V.R.F. Series,* No. 430, October 2007. Institute of Developing-Economies, Japan External Trade Organization.

[20] Sukono, Hidayat Y, Suhartono, Sutijo B, Bon A T B and Supian S 2016 Indonesian Financial Data Modeling and Forecasting By Using Econometrics Time Series and Neural Network. *Global Journal of Pure and Applied Mathematics*. ISSN 0973-1768 Volume 12, Number 4 (2016), pp. 3745-3757. © Research India Publications. http://www.ripublication.com/gjpam.htm.

[21] Wang X and Fu Y 2013 Some Characterizations of the Cobb-Douglas and CES Production Functions in Microeconomics. *Abstract and Applied Analysis*, pp. 1-6.

[22] Zushman J M 2016 Modelling greenhouse gas emissions from light duty vehicles. *30 ECTS thesis*, submitted in partial fulfillment of a *Magister Scientiarum* degree in Environment and Natural Resources, Faculty of Industrial Engineering, Mechanical Engineering and Computer Science University of Iceland.

IOP Conf. Series: Materials Science and Engineering **621** (2019) 012021 doi:10.1088/1757-899X/621/1/012021

New class of hybrid conjugate gradient coefficients with guaranteed descent and efficient line search

I.M. Sulaiman[1], Sukono[2*], Sudradjat Supian[2]and Mustafa Mamat[1*]

[1]Faculty of Informatics and Computing, Universiti Sultan Zainal Abidin, 21300, Terengganu, MALAYSIA

[2]Department of Mathematics' Faculty of Mathematics and Natural Sciences, UniversitasPadjadjaran, Bandung, INDONESIA.

Corresponding author: must@unisza.edu.my; sukono@unpad.ac.id

Abstract. Hybrid conjugate gradient (CG) techniques are one of the most prominent procedure for obtaining the solution of large-scale unconstrained optimization problems. This is due to its simplicity, global convergence, and low memory requirement. Numerous modifications have been done recently to improve the performance of these methods. In this paper, we proposed new class of hybrid CG coefficients with guaranteed descent under exact line search. Numerical results are presented to illustrate the efficiency of the proposed methodscompared to other classical CG coefficients.

1. Introduction

Conjugate gradient method (CG) is one of the robust line search methods designed to solve large scale unconstrained optimization problem [26]. The method has the following form

$$\min f(x) \text{ subject to } x \in R^n$$

where$f(x):\ R^n \to R$ is a continuously differentiable function. The CG methods are iterative methods of the form

$$x_{k+1} = x_k + \propto_k d_k \qquad (1)$$

where x_{k+1} is the current iterate, α_k is a step length obtained by exact line search, and d_{k+1} is the search direction computed as

$$d_{k+1} = \begin{cases} -g_{k+1} & for\ k = 0 \\ -g_{k+1} + \beta_{k+1} d_k & for\ k \geq 1 \end{cases} \qquad (2)$$

The scalar β_k known as the CG coefficient characterize different versions of the CG method based on their simplicity and global convergence. Some of the earliest and well-known formulas for the coefficient β_k are Hestenes and Stiefel [11] denoted as HS, Fletcher and Reeves [16, 17] denoted as FR, and Polak and Ribierre [3] denoted as PRP. These methods differ in terms of CG coefficient and performance and are defined as follows;

$$\beta_k^{FR} = \frac{\nabla f_k^T \nabla f_k}{\nabla f_{k-1}^T \nabla f_{k-1}}$$

$$\beta_k^{PR} = \frac{\nabla f_k^T (\nabla f_k - \nabla f_{k-1})}{\nabla f_{k-1}^T \nabla f_{k-1}}$$

$$\beta_k^{HS} = \frac{\nabla f_k^T (\nabla f_k - \nabla f_{k-1})}{d_{k-1}^T (\nabla f_k - \nabla f_{k-1})}$$

The global convergences of these coefficients have been studied by researchers in various field. For instance, the FR method was proved to be globally convergent by Zoutendijk [5] under exact line search. Citing the poor performance of this method due to jamming phenomenon, Powell [10] opposed the idea of Zoutendijk. Also, some researchers are of the opinion that PRP is the most reliable CG coefficient [12]. However, the PRP method might also not converge under exact line search and thus cycle indefinitely as shown by Powell [9]. For a sample research conducted to study the properties of CG methods, please refer to Al-Baali [8], Wei et al. [22], Touti-Ahmed and Storey [4], Gilbert and Nocedal [6], Hager and Zhang [18, 19], Rivaie et al. [12, 13],Abashar et al.[24], Yusoffet al.[25] and Dai and Yuan [21].For application of CG in estimating data, please refer to Shoid et al. [7].

Some hybrid version of the conjugate gradient parameters have been proposed such as β_k^{TS} by Touati-Ahmed and Storey [4] and β_k^{Hus} by Hu and Storey [20]. They combined two CG methods of PRP and FR to produce a new coefficient respectively as follows

$$\beta_k^{TS} = \begin{cases} \beta_K^{PRP}, & if\ 0 \le \beta_k^{PRP} \le \beta_k^{FR}, \\ \beta_k^{FR}, & \text{otherwise.} \end{cases}$$

$$\beta_k^{Hus} = max\left\{0, min\{\beta_k^{PRP}, \beta_k^{FR}\}\right\}$$

where$\nabla f_k$and$\nabla f_{k-1}$are the gradient of $f(x)$ at the point x_k and x_{k-1} respectively, and $\|.\|$ represents the Euclidean norm of vectors. The algorithms are globally convergence since the global convergence of the FR method and PRP methods have been proved. The algorithm also produces a better performance when compared to the FR, PRP and other CG coefficients.

In this paper, we present our new hybrid coefficientsβ_k^{ISM} and β_k^{ISM1} whose performance was compared with classical formulas of FR, PRP, SMAR, SMARZ, and SM coefficient. Section 2 present the new hybrid CG formula with its algorithm for solving unconstrained optimization problems. The sufficient descent proofs is presented in Section 3, followed by numerical illustrations and discussions in Section 4. The last section is the conclusion based on comparisons drawn from the results.

2. New proposed method

Recently, Sulaiman et al. [1, 2, 23] proposed some new nonlinear conjugate gradient formula known as SMAR (2015), SMARZ (2015), and SM (2018) and defined by

$$\beta_k^{SMAR} = \frac{\nabla f_k^T \left(\nabla f_k - \frac{\|\nabla f_k\|}{\|\nabla f_{k-1}\|} d_{k-1} \right)}{d_{k-1}^T d_{k-1}}$$

$$\beta_k^{SMARZ} = \frac{\nabla f_k^T \left(\nabla f_k - \frac{\|\nabla f_k\|}{\|\nabla f_{k-1}\|} d_{k-1} \right)}{d_{k-1}^T d_{k-1}}$$

$$\beta_k^{SM} = \frac{\nabla f_k^T \left(\nabla f_k - \frac{\|\nabla f_k\|}{\|\nabla f_{k-1}\|} d_{k-1} - d_{k-1} \right)}{d_{k-1}^T d_{k-1}}$$

Motivated by the efficiency and robustness of these methods, we propose some hybrid formula as follows;

$$\beta_k^{ISM} = max\left\{0, min\{\beta_k^{SMAR}, \beta_k^{SM}\}\right\} \tag{3}$$

$$\beta_k^{ISM1} = max\left\{\beta_k^{SMAR}, min\{\beta_k^{SMARZ}, \beta_k^{SM}\}\right\} \tag{4}$$

where ISM denotes Ibrahim Sulaiman, Mamat. The algorithm is given as follows:

Algorithm 1

Step 1: Initialization. Given $x_0 \in R^n$, set $k = 0$.

Step 2: Compute the CG coefficient β_k based on (3)and (4).

Step 3: Compute search direction d_k based on (2). If $g_k = 0$, then stop.

Step 4: Compute step size α_k based on exact line search.

Step 5: Update the new iterate base (1).

Step 6: Convergent test and stopping criteria.

If $f(x_{k+1}) < f(x_k)$ and $\|g_k\| \leq \varepsilon$ then stop.

Otherwise go to Step 2 with $k = k + 1$.

3. Sufficient descent condition

We need to show that β_k^{ISM}, andβ_k^{ISM1}satisfies the sufficient descent condition. For this to hold,

$$g_{k+1}^T d_{k+1} \leq -C\left\|g_{k+1}\right\|^2, \forall k \geq 0, C > 0 \tag{5}$$

The following Theorem shows that the proposed hybrid CG methods satisfies the sufficient descent condition with exact line search satisfies.

Theorem 1

Consider a CG method with search direction (2) and $\beta_k$given as (3 and 4), then condition (5) holds for all $k \geq 0$.

Proof.

We proof by induction. If $k = 0$, then $g_0^T d_0 \leq -C\|g_0\|^2$. Hence, condition (5) holds true.

Next, we want to show that for all $k \geq 0$, condition (5) will also hold true. By multiplying both sides of equation (2)by g_{k+1}^T, we get

$$g_{k+1}^T d_{k+1} = g_{k+1}^T\left(-g_{k+1}^T + \beta_{k+1} d_k\right) = -\left\|g_{k+1}\right\|^2 + \beta_{k+1} g_{k+1}^T d_k \tag{6}$$

However, we know that $g_{k+1}^T d_k = 0$ under exact line search [ ]. Hence,

$$g_{k+1}^T d_{k+1} = -\|g_{k+1}\|^2$$

Thus, the above condition holds true for $k + 1$. The proof is completed. ∎

4. Numerical Results

This section present the numerical results based on the comparisons with classical and existing methods such as FR, PRP, SMAR, SMARZ, and SM. All coefficients are implemented under exact line search procedure. The program was coded in MATLAB 7.6.0 (R 2013b) subroutine programming with stopping criteria $\|g_k\| \leq 10^{-6}$. The CPU processor used was Intel® corei5-2410M CPU @ 2.30 GHz processor, 4GB for RAM and Windows 7 Professional operating system [27]. The test functions considered as presented in Table 1 are from Andrei [15] with different dimensions $2 \leq n \leq 10000$ as suggested by Hillstrom [14]. To check the performance and global convergence of the proposed methods on the test problems, initial points closer and further away from the solutions were considered. This is done based on number of iterations and CPU time. The program is terminated whenever search procedure fails to find a positive step size and denote the algorithm as 'fail'.

Table 1. List of text functions.

No	Function	Dimension	Initial Points
1	Leon	2	(0.5,0.5), (3,3), (5,5), (-7,7)
2	Three Hump Camel	2	(0.5,0.5), (3,3), (5,5), (15,15)
3	Shallow	2	(2,2), (13,13), (23,23), (27,27)
4	Dixon and Price	2,4	(1,1), (15,15), (25,25), (35,35)
5	Ext White & Holst	2,4,10	(0.5,0.5,…,0.5), (3,3,…,3), (5,5,…,5), (15,…,15)
6	Extended Beale	2,4,10,100	(2,2,…,2), (3,3,…,), (5,5,…,5), (10,10,…,10)
7	ExtQuadratic Penalty	2,4,10,100	(3,3,…,3), (11,11,…,11), (13,13,…,13), (25,25,…25)

8	Extended Quartic	2,4,10,100, 1000,10000	(5,5,…,5), (10,10,…,10), (15,15,…,15), (20,20,…,20)
9	Rosenbrook	2,4,10,100, 1000,10000	(10,10,…,10), (13,13,…,13), (23,23,…,23), (31,31,…,31)

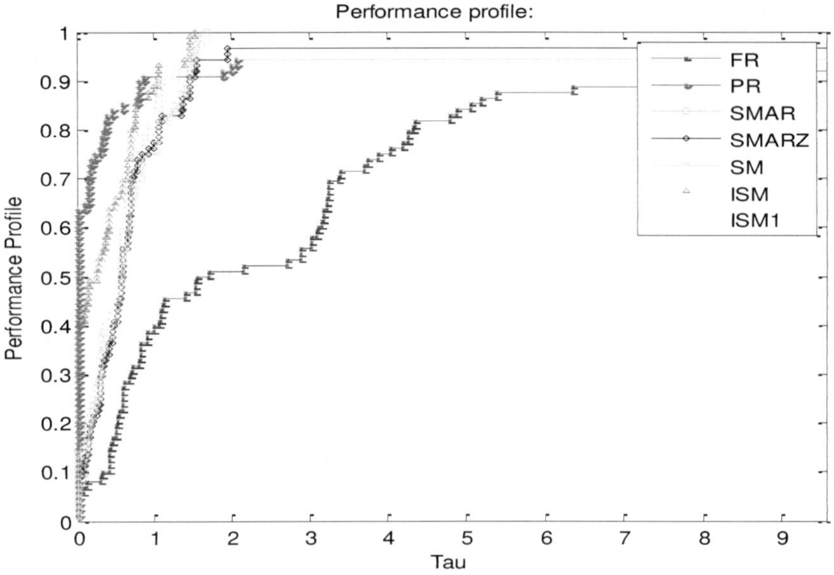

Figure 1. Performance profile based on number of iterations

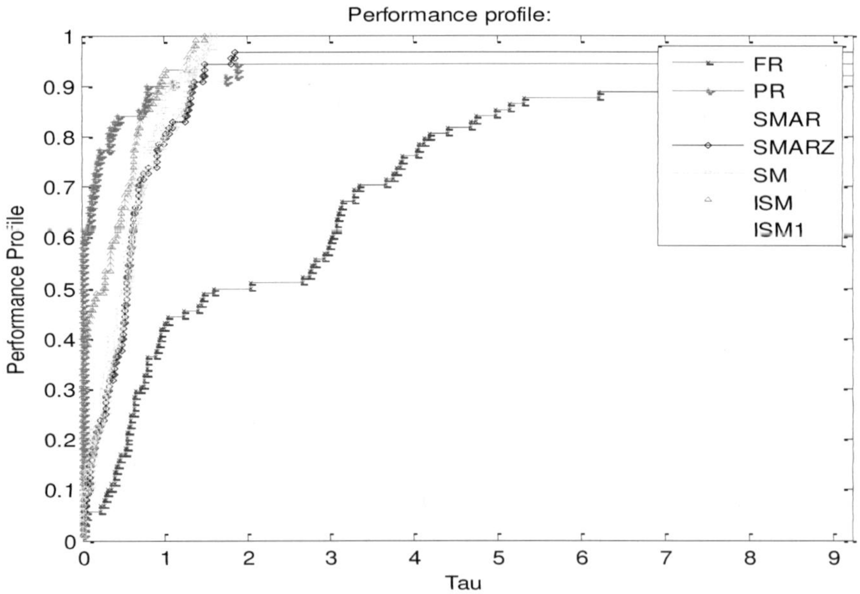

Figure 2. Performance profile based on CPU time

Figure 1and Figure 2 above illustrates thenumerical performance of the proposed class of hybrid CG coefficients with comparison to the classical CG coefficients of FR, PRP, SMAR, SMARZ, and SM. This is based on number of iterations and CPU time. The Figures also shows that the proposed coefficients ISM and ISM1 outperformed the classical methods in terms of number of iteration and CPU time. Also, based on overall performance analysis, refer to Table 2 below.

Table 2. Overall performance comparison.

Methods	FR	PRP	SMAR	SMARZ	SM	ISM	ISM1
Percentage (%)	92.0455	94.3182	100.0000	96.5909	98.8636	100.0000	100.0000
Success (%)	5.6818	63.6364	13.6364	9.0909	9.0909	39.7727	13.6364
No of Fail	7.0000	5.0000	0	3.0000	1.0000	0	0

Though, the classical CG coefficients of FR and PRP are regarded the best CG methods, and SMAR, SMARZ, and SM are some of the recent methods with robust performance but areunable to solve of all the test problems. Whereas, the new class of hybrid CG method of ISM and ISM1 solve all the test functions and reach 100%. Hence, the proposed methods are considered superior compare to the other methods.

5. Conclusions

The hybrid CG techniques plays important role in solving large scale unconstrained optimization problems. Numerous studies have been done on this method due to its simplicity, and efficient convergence properties. In this paper, we presentednew and simple class of hybrid CG coefficients known as ISM and ISM1 with sufficient descent condition under exact line search. Numerical computations have shown that the proposed methods perform better that the classical methods of FR, PRP, SMAR, SMARZ and SM both in terms of number of iteration and CPU time.However, further research in this area can test the performance of these new coefficients under other line search procedures.

References

[1] Ibrahim S M, Mamat M, Abdelrhaman A, Rivaie M, Fatma S M and Zabidin S 2014 The Global Convergence Properties of an Improved Conjugate Gradient Method, *Applied Mathematical Sciences*, **9**, 1857-1868.

[2] Ibrahim S M, Mamat M, Abdelrhaman A, Rivaie M and Zabidin S 2015 A Modified Nonlinear Conjugate Gradient Method for Unconstrained Optimization. *Applied Mathematical Sciences,* **9**, 2671 - 2682.

[3] Polyak B T 1969 The conjugate gradient method in extreme problems, *USSR Comp. Math. Phys.* **9**, 94-112.

[4] Touati-Ahmed D and Storey C 1990 Globally convergent hybrid conjugate gradient methods, *J.Optim.Theory Appl.***64,** 379-397.

[5] Zoutendijk G 1970 Nonlinear programming computational methods, in: J. Abadie, (Ed.), *Integer and Nonlinear Programming, North- Holland, Amsterdam*, 37-86.

[6] Gilbert J C, Nocedal J 1992 Global convergence properties of conjugate gradient methods for optimization,*SIAM. J. Optim.* **2**, 21-42.

[7] Shoid S, Shapiee N, Zull N, Abdul Ghani N H, Mohamed N S, Rivaie M and Mamat M 2018 The application of new conjugate gradient methods in estimation data. *Int. J. of Engrand Tech.* **7**(2.14), 25-27.

[8] Al-Baali M 1985 Descent property and global convergence of the Fletcher-Reeves method with inexact line search, *IMA J. Numer. Anal.***5**, 121-124.

[9] Powell M J D 1977 Restart procedures for the conjugate gradient method, *Math. Program.***12**, 241-254.

[10] Powell M J D 1977 Nonconvex minimization calculations and the conjugate gradient method in Lecture notes in mathematics, 1066,*Springer-Verlag, Berlin*, 122–241.

[11] Hestenes M R and Stiefel E L 1952 Methods of conjugate gradients for solving linear systems, *J. Research Nat. Bur. Standards,***49**, 409-436.

[12] Rivaie M, Mamat M, Leong W J and Ismail M 2012 A new class of nonlinear conjugate gradientcoefficients with global convergence properties, *Appl. Math. Comput.* **218**, 11323-11332.

[13] Rivaie M, Abashar A, Mamat M and Mohd I 2014 The convergence properties of a new type of conjugate gradient methods, *Appl. Math. Sci*,**8**, 33-44.

[14] Hilstrom K E 1977 A simulation test approach to the evaluation of nonlinear optimizationalgorithms, *Acm. Trans. Math. Softw.* **3**, 305-315.

[15] Andrei N 2008 An unconstrained optimization test functions collection, *Adv. Modell. Optim.* **10**, 147-161.

[16] Fletcher R and Reeves C 1964 Function minimization by conjugate gradients, *Comput. J.***7**, 149-154.

[17] Fletcher R 1987 Practical Method of Optimization,*Unconstrained Optimization*.Vol. I,Wiley, New York.

[18] Hager W W and Zhang H C 2005 A new conjugate gradient method with guaranteed descent and efficient line search, *SIAM J. Optim.* **16**, 170-192.

[19] Hager W W and Zhang H C 2006 A survey of nonlinear conjugate gradient methods, *Pacific Journal of Optimization***2**(1), 335-58.

[20] Hu Y F and Storey C 1991 Global convergence result for conjugate gradient methods. *J. Optim. Theory. Appl.*, **71**, 399-405

[21] Dai Y H and Yuan Y 2000 A nonlinear conjugate gradient method with a strong globalconvergence property, *SIAM J. Optim.***10**, 177–182.

[22] Wei Z, Yao S and Liu L 2006 The convergence properties of some new conjugate gradientmethods, *Appl. Math. Comput.* **183**, 1341-1350.

[23] Ibrahim S M and Mamat M 2018 A new CG solution technique for unconstrained optimization. *Proc. of the 2nd IEEE Int. Conf. on Intelligent Systems Engineering* (Kuala Lumpur), Malaysia.

[24] Abashar A, Mamat M, Rivaie M, Mohd I, Omer O 2014 The proof of sufficient descent condition for a new type of conjugate gradient methods, *AIP Conf Proceeding*, 1602, 296-303.

[25] Yusoff N H M, Mamat M, Rivaie M, Mohd I 2014 A new conjugate method for unconstrained optimization with sufficient descent, AIPConf Proceeding, 1602, 514-519.

[26] Zull N, Aini N, Mamat M, Ghani N H A, Mohamed N S, Rivaie M and Sukono 2017 An Alternative Approach for Conjugate Gradient Method. *Journal of Engineering and Applied Sciences,* 12 (18), 4622-4626, 2017. ISSN: 1816-949X. ©Medwell Journals, 2017.

[27] Napitupulu H, Sukono, Mohd I B, Y Hidayat Y and Supian S 2018 *IOP Conf. Series: Materials Science and Engineering*, 332 (2018), 012024. doi:10.1088/1757-899X/332/1/012024.

AUTHOR INDEX

Institute of Physics
Dirac House, Temple Back
Bristol BS1 6BE UK

ISSN: 1757-8981
ISBN 978-1-7138-0901-2

9 781713 809012

CURRAN ASSOCIATES INC.
proceedings
.com